AI and Renewable Energy Storage Synergy

Renewable energy generation is crucial for sustainability but its inconsistency in availability requires better storage solutions. The fusion of renewable energy technologies with AI can yield positive outcomes, such as increased energy efficiency, reduced carbon emissions, improved grid stability, and optimized storage techniques and methods. This book provides a state-of-the-art perspective on thermal energy storage (TES) technologies integrated with renewable energy technologies and Artificial Intelligence (AI). It includes various case studies and practical applications that offer actionable insights for professionals and researchers in the field and addresses real-world challenges.

Features

- Provides a comprehensive overview of thermal energy storage methods and systems.
- Explains AI's role in enhancing thermal energy storage.
- Includes real-world case studies and discusses perspectives of how TES and AI contribute to sustainability.
- Addresses the integration of AI with digital twin technology, showcasing how AI algorithms can enhance the predictive accuracy and operational efficiency of digital twins in TES applications.
- Explores various AI-driven strategies and technologies that enhance the efficiency, reliability, and scalability of energy storage systems.

This book is for researchers, academics, and graduate students in Energy and Environmental Sciences, and those interested in renewable energy, AI, and energy storage. It's also an excellent reference for industry and government professionals, energy policy makers, and analysts.

Sustainable Engineering and Science

Series Editor: Dr. Humbul Suleman
Senior Lecturer in Chemical Engineering Teesside University, UK

The *Sustainable Engineering and Science* book series integrates industrial, social, and environmental processes in a global context, requiring a triangular approach that includes social value, environmental impacts, and economic value. This series explores the cutting-edge research, innovations, and practices that contribute to building a resilient yet environmentally conscious society. It provides a comprehensive and accessible resource on a wide range of topics related to sustainability from an engineering and scientific perspective.

Membrane Technology for Sustainability: Applications in Gas and Water Treatment
Rizwan Nasir, Danial Qadir, and Hafiz Abdul Mannan

AI and Renewable Energy Storage Synergy: Fundamentals and Applications
Edited by Naveed Ahmed and Mumtaz A. Qaisrani

For more information about this series, please visit: www.routledge.com/Sustainable-Engineering-and-Science/book-series/CRCSSE

AI and Renewable Energy Storage Synergy
Fundamentals and Applications

Edited by
Naveed Ahmed and Mumtaz A. Qaisrani

CRC Press
Taylor & Francis Group
Boca Raton London New York

CRC Press is an imprint of the
Taylor & Francis Group, an **informa** business

Designed cover image: © Getty Images

First edition published 2027
by CRC Press
2385 NW Executive Center Drive, Suite 320, Boca Raton FL 33431

and by CRC Press
4 Park Square, Milton Park, Abingdon, Oxon, OX14 4RN

CRC Press is an imprint of Taylor & Francis Group, LLC

ISBN: 978-1-032-85018-4 (hbk)
ISBN: 978-1-032-85019-1 (pbk)
ISBN: 978-1-003-51613-2 (ebk)

DOI: 10.1201/9781003516132

Typeset in Palatino
by KnowledgeWorks Global Ltd.

This book is dedicated to our parents.

Contents

Preface

The global energy sector is undergoing a profound transformation driven by the urgent need to reduce carbon emissions, enhance energy security, and meet rapidly growing energy demands. Renewable energy technologies such as solar, wind, and geothermal have emerged as key pillars of this transition. However, their inherent intermittency and variability pose significant technical and operational challenges. Energy storage systems—particularly thermal energy storage—play a critical role in bridging the gap between energy generation and demand, enabling reliable, efficient, and flexible energy systems.

In parallel, artificial intelligence (AI) has rapidly evolved as a powerful enabler across engineering and energy disciplines. Advances in machine learning, data-driven modeling, optimization, and digital twin technologies have opened new pathways for improving the design, operation, and management of energy storage systems. When combined with renewable energy and storage technologies, AI offers unprecedented opportunities to enhance system performance, reduce costs, extend operational lifetimes, and support intelligent decision-making in complex energy environments.

This book, *AI and Renewable Energy Storage Synergy: Fundamentals and Applications*, is conceived to address this emerging convergence of renewable energy, energy storage, and artificial intelligence. It provides a comprehensive and structured exploration of fundamental concepts, enabling technologies, modeling approaches, and real-world applications. The volume covers a broad spectrum of topics, ranging from renewable energy systems and thermal energy storage technologies to AI-driven optimization, smart energy management systems, digital twin solutions, and future research directions.

The chapters are authored by experts from academia and industry and are designed to balance theoretical foundations with practical insights. Particular emphasis is placed on thermal energy storage systems across low-, medium-, and high-temperature ranges, reflecting their growing importance in buildings, industrial processes, and power generation. Through modeling frameworks, case studies, and applied examples, the book demonstrates how AI can transform conventional energy storage systems into intelligent, adaptive, and sustainable solutions.

This book is intended for researchers, graduate students, engineers, and professionals working in renewable energy, energy storage, and smart energy systems. It also serves as a valuable reference for policymakers and practitioners seeking to understand the role of AI in advancing sustainable energy technologies. By presenting both current developments and future perspectives, this volume aims to contribute meaningfully to the ongoing global transition toward resilient, low-carbon, and intelligent energy systems.

Naveed Ahmed
Mumtaz Ahmad Qaisrani

About the Editors

Naveed Ahmed is an Associate Professor at the US–Pakistan Center for Advanced Studies in Energy, National University of Sciences and Technology (NUST), Islamabad. He is an accomplished researcher in energy technologies with over 42 peer-reviewed journal publications, several book chapters, and a strong record of externally funded research projects. He has actively contributed to leading international conferences in the field of energy systems. With more than 10 years of combined postdoctoral and industrial experience, Dr. Ahmed has established himself as a recognized academic contributing to advancements in renewable energy and thermal energy storage research.

Mumtaz A. Qaisrani, a Tenured Assistant Professor, specializes in Energy Engineering with a focus on renewable energy, solar and wind systems, and energy efficiency. He has published over 30 papers in SCI journals in his field, and he has reviewed over 40 manuscripts for various publishers. As an Academic Editor for the *International Journal of Energy Research* (IF 4.6), he contributes to the scientific community. His diverse academic background and extensive experience drive holistic research advancements in Energy Engineering.

Contributors

Muhammad Farasat Abbas
Department of Electrical and Photonics Engineering
Technical University of Denmark
Kongens, Lyngby, Denmark

M. Asim Amin
Southern University of Science and Technology
China
and
Rapid Volt (PVT) Ltd.
Pakistan

Ayesha Arif
National University of Sciences and Technology
Islamabad, Pakistan

Bilawal A. Bhayo
Department of Mechanical Engineering
Mehran University of Engineering and Technology
Shaheed Zulfiqar Ali Bhutto Campus Khairpur Mir's
Khairpur, Pakistan

Karem Elsayed Elfeky
College of Power and Energy Engineering
Harbin Engineering University
Harbin, China
and
Department of Mechanical Engineering
Benha Faculty of Engineering
Benha University
Banha, Egypt

Mohammad Hamid
Department of Energy and Power Engineering
School of Mechanical Engineering
Beijing Institute of Technology
Beijing, China

Muhammad Ihsan-Ul-Haq
Materials Science and Engineering Department
King Fahd University of Petroleum and Minerals
Dhahran, Saudi Arabia
and
Interdisciplinary Research Center for Sustainable Energy Systems
King Fahd University of Petroleum and Minerals
Dhahran, Saudi Arabia

Ahmad Jamil
Department of Energy and Petroleum Engineering
University of Stavanger
Stavanger, Norway

Youwei Jia
Southern University of Science and Technology
China

Yabin Jin
School of Energy and Architecture
Xihang University
Xi'an, China

Keke Shao
Department of Energy and Power Engineering
School of Mechanical Engineering
Beijing Institute of Technology
Beijing, China

Atif Naveed Khan
US Pakistan Centre for Advanced Studies in Energy (USPCAS-E)
National University of Sciences and Technology (NUST)
Islamabad, Pakistan

Naseer Ahmad Khan
US Pakistan Centre for Advanced Studies in Energy (USPCAS-E)
National University of Sciences & Technology (NUST)
Islamabad, Pakistan

Laveet Kumar
Department of Mechanical Engineering
Mehran University of Engineering and Technology
Jamshoro, Pakistan

Yonghui Liang
Department of Energy and Power Engineering
School of Mechanical Engineering
Beijing Institute of Technology
Beijing, China

Veerta Mah-i-Liqua
The Islamia University of Bahawalpur
Bahawalpur, Punjab, Pakistan

Song Mengjie
Department of Energy and Power Engineering
School of Mechanical Engineering
Beijing Institute of Technology
Beijing, China

Saad Naseem
College of Computing
Georgia Institute of Technology
Atlanta, GA, USA

and

The Joint BioEnergy Institute
Lawrence Berkeley National Laboratory
Emeryville, CA, USA

and

Biological Systems and Engineering Division
Lawrence Berkeley National Laboratory
Berkeley, CA, USA

Shahryar
US Pakistan Centre for Advanced Studies in Energy (USPCAS-E)
National University of Sciences and Technology (NUST)
Islamabad, Pakistan

Aye Aye Thet
Department of Energy and Power Engineering
School of Mechanical Engineering
Beijing Institute of Technology
Beijing, China

Ameer Hamza Tiwana
National University of Sciences and Technology
Islamabad, Pakistan

Muhammad Waseem
Maynooth University
Ireland

Yasir Yasin
Khwaja Fareed University of Engineering and Information Technology
Rahim Yar Khan, Punjab, Pakistan

Sirui Yu
Department of Energy and Power Engineering
School of Mechanical Engineering
Beijing Institute of Technology
Beijing, China

Acknowledgements

We gratefully acknowledge the support of our respective institutions throughout the extended period required to compile this book, particularly the Department of Thermal Energy Engineering at the U.S.–Pakistan Center for Advanced Studies in Energy, National University of Sciences and Technology (NUST), and Institute of Mechanical and Manufacturing Engineering, Khwaja Fareed University of Engineering and Information Technology. We also acknowledge the support of the University of Stavanger and University College Dublin, Ireland, during the early stages when the idea for this book was conceived. Furthermore, we sincerely appreciate the valuable contributions of the reviewers, whose constructive feedback made this work possible through the review process.

Nomenclatures and Symbols

Variable	Description
ΔH_1	Specific latent heat of the material
ΔU	Change in Internal Energy
η	Overall Efficiency of Renewable Energy System
AI	Artificial Intelligence
ANN	Artificial Neural Network
BOG	Boil Off Gas
CES	Cold energy storage
CFD	Computational Fluid Dynamics
CHP	Combined Heat and Power
COP	Coefficient of Performance
c_p	Specific heat capacity
CSP	Concentrated Solar Power
CTES	Cold Thermal Energy Storage
DH	District Heating
DL	Deep Learning
DT	Digital Twin
EMS	Energy Management System
ESS	Energy Storage System
GA	Generic Algorithms
HDPE	High-Density Polyethylene
HTF	Heat Transfer Fluid
HVAC	Heating, Ventilation, and Air Conditioning
LAES	Liquid Air Energy System
LHS	Latent Heat Storage
LNG	Liquified Natural Gas
LS-SVM	Least Square Support Vector Machine
LTTS	Long-Term Thermal Storage
m	Mass
ML	Machine Learning
PCMs	Phase Change Materials
PCT	Phase Change Temperature
PP	Polypropylene
PSO	Particle Swarm Optimization
PTES	Pumped Thermal Energy Storage
PV	Photovoltaic
Q	Heat
Q_{input}	Heat Input to the System
Q_s	Sensible thermal energy stored

RET Renewable Energy Technology
SFES Superconducting Flywheel Energy Storage
SHS Sensible Heat Storage
STES Seasonal Thermal Energy Storage
STTS Short Term Thermal Storage
T Temperature
TES Thermal Energy Storage
TESS Thermal Energy Storage System
W Work
W_{useful} Amount of Useful Work Produced
x_1 Proportion of the storage material that underwent a phase change

1

Introduction and Overview of Current Energy Challenges and the Transition to Renewable Energy Sources

Shahryar, Atif Naveed Khan, and Muhammad Farasat Abbas

1.1 Introduction: The Urgency of Energy Transition

An exceptional energy system change takes place globally because climate change, energy security, and reasonable economic development problems need immediate solutions. The progressing 21st century brings forward a two-fold challenge to the energy sector regarding the continued growth of global energy consumption and the reduction of environmental and social effects from fossil fuel dependence, which is clearly shown in Figure 1.1. The energy transformation toward sustainable assets, coupled with environmental stability, represents an obligatory path for maintaining sustainable planetary stability while creating economic success [1].

Energy consumption has steadily increased due to the quick industrial development of emerging nations, along with urban growth, digital transformation, and worldwide standards of living improvements. Traditional energy systems, which produce carbon emissions, serve as the main human-made cause of climate change. According to the Intergovernmental Panel on Climate Change (IPCC), global carbon dioxide (CO_2) emissions must drop to approximately 55% below their 2010 values by 2030 before becoming zero around 2050 for a 1.5°C maximum warming limit. The realization of this reduction demands comprehensive rethinking throughout every stage of the power production and delivery systems, together with distribution, management, and consumption methods [2].

The initial portion of this text establishes the necessary groundwork to analyze artificial intelligence (AI) and renewable energy storage systems that can work together to boost energy transformations. The need for forward-thinking integrated energy system designs is explained by the global energy inconsistency as well as the key role of energy in economics and nature alongside the recent technological revolution of infrastructure systems.

DOI: 10.1201/9781003516132-1

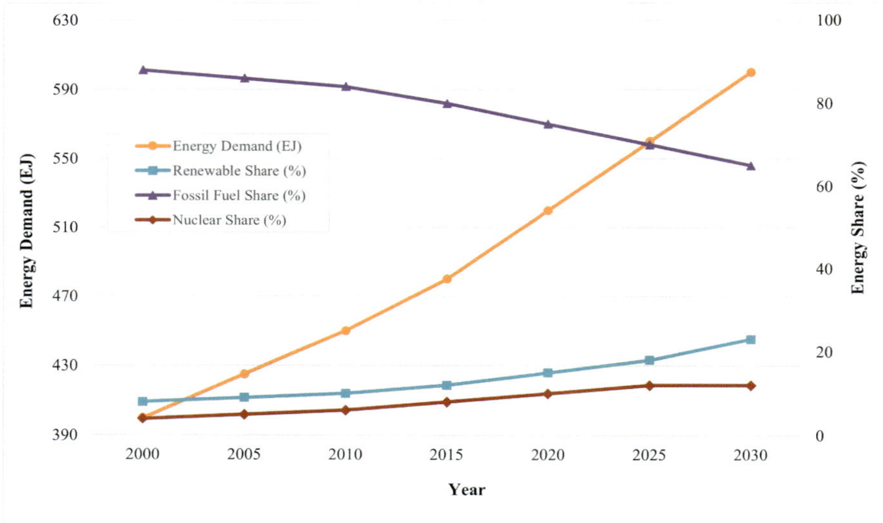

FIGURE 1.1
Global energy demand and source shares (2000–2030). (*Source*: IEA World Outlook, 2023.)

1.1.1 The Global Energy Inconsistency: Rising Demand vs. Sustainability Needs

The global energy paradox exists as an opposing force between mounting energy requirements and immediate system decarbonization needs which protect the environment from disaster. The International Energy Agency (IEA) predicts primary energy needs worldwide will expand by 1.3% per year between 2019 and 2040 while electric power requirements will surge at 2.5% annually per year during the same period according to World Energy Outlook 2023. Non-OECD countries represent the main drivers behind projected energy consumption growth since they will use energy to reduce poverty while accelerating industrial development and building infrastructure for urbanization needs as seen in Table 1.1 on the next page.

Over a year of intense renewable energy expansion has not been enough to offset the 82% dominance of fossil fuel energy usage including coal, oil, and natural gas in global consumption during 2022. The total greenhouse gas (GHG) emissions originating from these sources surpass 75%, while coal continues to be the leading cause of CO_2 emissions from energy consumption. The world depends on fossil fuels for power mainly because aging pipelines need replacement and monetary systems, and government programs have limited impact.

A significant number of over 770 million people mainly residing in sub-Saharan Africa and parts of Asia experience electricity shortages while

TABLE 1.1
Global Energy Demand vs. Renewable Energy Share (2000–2025)

Year	Global Primary Energy Demand (EJ)	Renewable Energy Share (%)	Estimated Annual Growth (Demand, %)	Data Source/Reference
2000	400	8	–	IEA (World Energy Outlook 2023)
2005	425	9	~1.2% over 5 years	IEA (Historical Data)
2010	450	10	~1.2% over 5 years	IEA/BP Statistical Review
2015	480	12	~1.3% over 5 years	IEA (World Energy Outlook 2023)
2020	520	15	~1.6% over 5 years	IEA/World Energy Outlook 2023
2025*	560	18	~1.5% over 5 years (Projected)	IEA Projections

billions more experience energy instability. The global energy system needs to handle three interconnected tasks: increase accessibility, preserve reliability, and achieve substantial emission reductions in an extremely complex manner.

Sustainability is the context of extending past environmental concerns. There are three major factors involved:

- In environmental sustainability, all operations must reduce pollution through strategies aimed at resource management and protection of ecosystems.
- Sustainable economics requires obtainable and economical energy access, resilient infrastructure, and financial stability through lasting periods.
- The social sustainability framework should provide simple access to energy resources and empower clean energy solutions that facilitate community empowerment.

To resolve this paradox, companies must deliver radical changes throughout their entire value chain, which need to start at production levels and end-to-end use stages and policy implementation. There are multiple key enablers involved in decentralized renewable energy systems, which are most relevant to AI-powered decision support tools and digital energy platforms helping to resolve the problems.

1.1.2 The Role of Energy in Economic Growth and Climate Action

Energy cannot be taken as a commodity; it is an engine of economic development, social advancement, and technological innovation. Historical economic expansion developed a direct relationship with increased energy consumption because energy powers industrial and service activities and transportation and nationwide productivity. The established relationship between economic growth and energy consumption which results in additional emissions remains under review because of damaging environmental consequences.

The current challenge focuses on disentangling economic development from carbon pollution, while new strategies involving renewable power generation and technology improvements and energy digitalization search for higher accessibility. The concept of 'green growth' supports economic progress that meets two essential criteria: it must be inclusive, benefiting all segments of society, and it must foster environmental sustainability by creating conditions that restore and regenerate natural ecosystems.

1.1.2.1 Energy Intensity and Decarbonization

Energy intensity represents a significant measurement factor that determines energy use per GDP unit. Global energy intensity showed a yearly improvement rate of 1.3% between 2010 and 2020, although annual goals of 3% improvement are required to achieve net-zero targets. Relatively decoupled advanced economies merge with emerging economies that prefer to bypass conventional fossil fuels through the use of renewable energy systems [3].

Renewable energy sources deliver two main advantages for the economy. The energy sector offers low-emission power solutions at competitive prices to substitute for fossil fuel usage. The second positive aspect of renewable energy creates entirely new economic sectors that boost supply networks and generate workforce positions. The renewable energy sector worldwide hired more than 13 million people during 2022, while industries will observe massive employment growth during the future expansion of solar power and wind power in addition to green hydrogen and storage systems [4].

1.1.2.2 Energy and Climate Integration

The effort to minimize climate change depends heavily on the energy sector. Low-carbon technologies along with carbon pricing and electricity-powered transport systems and improved power grid infrastructure form essential components in climate-related national and international plans. The IPCC Sixth Assessment Report (2021–2022) lists energy system transformation among the most powerful measures to decrease emission levels.

Energy serves as a fundamental element to reaching various United Nations Sustainable Development Goals (SDGs) [5].

SDG 7: Affordable and Clean Energy

SDG 9: Industry, Innovation, and Infrastructure

SDG 13: Climate Action

The achievement of a fair, inclusive energy transition depends on specific policy measures that help vulnerable communities alongside worker transition and private and public energy sector development.

1.1.3 Disruptive Technologies Shaping the Future of Energy

Modern times have brought about quick digital transformation and technological unification in the global energy industry through disruptive technologies which transform how energy is generated, transmitted, and consumed.

1.1.3.1 Artificial Intelligence and Machine Learning

The approach to managing energy is transforming due to AI technologies that make systems perform more intelligently. Predictive maintenance along with real-time monitoring forms two of its key uses that also include smarter pricing and better decision-making through data analysis. There are key applications included in it:

- Deep learning forecasting algorithms help demonstrate an accurate view of future electricity requirements through precise time-based assessment.
- The prediction analysis of renewable energy systems focuses specifically on wind and solar farms to address operational obstacles created by variable power generation.
- The power grid management system called Distribution Management System uses AI tools to operate efficiently while requiring reduced human operator involvement.
- Devices linked to non-intrusive load monitoring (NILM) together with machine vision methods separate energy consumption into specific appliance levels.

Including AI technology in SCADA systems together with energy management platforms enables utility companies to conduct rapid response actions toward demand shifts and system faults ahead of their emergence. The key components are maintaining the lead while maintaining continuous system operation.

1.1.3.2 Smart Grids and Internet of Energy (IoE)

The main components of a smart grid system include sensors as well as control systems and two-way communication networks. The system helps users perform decentralized adaptive energy management in a controlled environment. These systems allow:

- The implementation of distributed energy resources (DERs) through rooftop solar together with micro-wind and electric vehicles receives integration.

- The implementation of price signals in real time allows customers to shift their energy consumption while controlling peak demand levels.
- The system maintains greater operational strength because it can identify failures while separating into independent operation areas and rerouting electricity automatically.

IoE implements a distributed network based on IoT infrastructure to unite billions of smart devices that exchange information between various energy sectors. The digital mesh created by this structure serves as the foundation for managing demand flexibility, performing advanced smart meter collection, and making predictions about infrastructure maintenance needs.

1.1.3.3 Energy Storage and Thermal Energy Storage (TES)

Interruption in power supply stands as one of the main difficulties when relying on renewable energy systems. Energy storage solutions are required to capture excess energy from solar or wind generation systems because they cannot redistribute this power effectively. Recent advances in:

- Lithium-ion batteries (increased energy density, longer cycle life)
- Solid-state batteries (enhanced safety, higher-voltage thresholds)
- Flow batteries (scalable and durable for grid applications)

Thermal energy storage (TES) implements three main technologies which incorporate molten salts, phase change materials, and sensible heat storage devices for cost-efficient TES applications during power generation and industrial manufacturing operations. The TES technology provides optimal performance for CSP-based facilities alongside district heating networks since it efficiently stores large amounts of thermal energy without significant conversion inefficiencies.

Such technologies enable the development of a versatile integrated power system. These Technologies can adjust supply and demand dynamics to achieve better energy system stability.

1.2 Current Energy Challenges and Barriers to Sustainability

The rise in clean energy support does not eliminate the fundamental obstacles that block global energy transition including structural, environmental, economic, and geopolitical factors as described in Table 1.2 on the next page. These obstacles consist of technical aspects and socio-political elements as

TABLE 1.2

Summary of Current Global Energy Barriers

Challenge Area	Description	Example Region/Case	Impact Metric/ Magnitude	Proposed Mitigation Strategies	Reference
Climate and emissions	Fossil fuel combustion generating high CO_2 emissions	Global; industrialized economies	~36.8 Gt CO_2 (2023 estimate)	Decarbonization, CCS, renewable integration	IPCC, IEA (2023)
Economic volatility	Price shocks and market instability due to fossil fuel dependence	Europe (e.g., Russia–Ukraine conflict)	Gas price spike >250% (2022 crisis)	Diversification, renewable investments, strategic reserves	IEA, World Bank Reports
Energy poverty	Lack of access to modern energy in underserved regions	Sub-Saharan Africa	~600 million people without electricity	Off-grid solar, mini-grids, micro-hydro projects	UN SDG Report, IEA
Infrastructure limitations	Aging grid systems unable to integrate distributed renewables	U.S., India	15–30% transmission & distribution losses	Grid modernization, smart grid tech, energy storage solutions	National Grid Reports, IEA
Policy and regulatory gaps	Inconsistent policies slowing renewables deployment	Various (policy delays)	Low renewable adoption rates; lengthy permitting times	Stable policy frameworks, carbon pricing, streamlined permitting processes	IEA, OECD Energy Policy Reviews

well as market dynamics. The increasing pressure for climate mitigation requires immediate evaluation of multiple risks that limit sustainable energy system implementation. The section describes the principal obstacles that hinder the worldwide energy transformation by classifying them into environmental-climate risks and economic-geopolitical complications as well as systemic structural limitations.

1.2.1 Environmental and Climate Risks

Climate change alongside more frequent extreme weather patterns has demonstrated that existing energy system environmental risks exist across the current time frame. The manufacturing of energy using fossil fuels generates substantial environmental consequences that create worldwide and immediate effects.

1.2.1.1 Global Carbon Footprint and Emissions Crisis

Energy frameworks currently generate the biggest share of human-made GHG emissions since coal and petroleum, and natural gas combustion remain widespread. Despite the fast expansion of renewable energy sources, the total energy-related CO_2 emissions reached 36.8 billion metric tons and set another historical record during 2022. Current climate change has elevated global temperatures by 1.1° Cover pre-industrial periods because of these primary emissions [6].

The projected climate path without meaningful decarbonization activities will surpass two degrees Celsius of warming during 2100 thus initiating permanent climate change effects like polar ice disappearance and sea level increase along with permafrost melting. Methane escapes during the operation of fossil fuel infrastructure despite getting less attention which poses another major threat. Energy systems currently overlook the high global warming impact of methane because its 20-year GWP exceeds CO_2 by more than 80 times.

Energy sector entities need to decrease emissions within all parts of their operational value chains that cover direct emissions from source one and indirect emissions from source two together with emissions from sold products through source three initiatives. A combination of accurate emissions reporting standards together with strict imposition of carbon pricing requires proper implementation of decarbonization strategies. The situation will remain unchanged unless strong actions are taken.

1.2.1.2 Ecological and Public Health Impacts

The use of fossil fuels generates extensive ecological deterioration as well as health problems for humans. Winning fossil fuels through extraction methods damages biodiversity as well as contaminates freshwater supplies while striking a detrimental blow to land health.

Initial power plant releases and transport-based emissions have been scientifically linked to heart problems, disease-based respiratory conditions, and higher mortality rates. The World Health Organization reports that air pollution leads to seven million yearly fatalities mainly through energy combustion processes [7].

Power plant thermal pollution along with waste heat contamination negatively impacts aquatic environments. The persistent release of oil and deposition of coal ash materials generate harmful ecological effects. The adoption of renewable energy systems decreases environmental impacts though life cycle assessments should be made to stop new ecological issues that arise from extracting raw materials, especially lithium and cobalt, needed for battery production.

1.2.2 Economic and Geopolitical Risks

Environmental considerations play only a partial role in driving the global energy transition because economic and geopolitical determinants make the current energy situation complex. Monetary and political concerns from fossil fuel-based energy systems continue to generate global stability, economic, and access to energy issues. The following list contains the most significant problems in this sector.

1.2.2.1 Fossil Fuel Dependence and Market Volatility

Modern economies today face marketplace instabilities because they remain tightly dependent on fossil fuel resources. The natural tendency for oil and gas market prices to be unstable occurs because of political disruptions and also because of OPEC+ members' production controls and speculative activities.

Natural gas prices soared more than 300% around the world because of Russia-Ukraine geopolitical tensions this year causing European regions to face significant energy supply issues alongside rising inflation and resource restrictions during the 2022 energy crisis. The two factors that negatively affected developing economies were their heavy dependence on imported fuels and minimal financial reserves. Renewable investment faces deregulation because of $1 trillion global fossil fuel subsidies which lower market prices artificially during 2022. The economic rationale to eliminate subsidies aligns with political challenges because their abolition raises fuel costs in regions that depend on price subsidies.

1.2.2.2 Energy Access, Equity, and Affordability

The worldwide advancement has failed to eliminate the widespread existence of energy inequality. The lack of electricity access affects more than 770 million people while above 2.4 billion people use traditional biomass for cooking because this method creates safety and health risks primarily for women and children.

The lack of energy access hinders educational progress and restricts hospital service provision and local business expansion. Energy costs represent an excessive financial burden on households that fall below the national income average in developed countries as well as developing nations. Increase in electricity rates caused by global market movements and decarbonization expenses may increase social inequality because appropriate support systems are not in place.

UN SDG 7 requires that all human beings have access to affordable, reliable, sustainable, modern energy which defines both a basic human right and a fundamental condition for inclusive development. The combination of off-grid solar, mini-grids, and clean cooking technologies appears as an effective solution although they struggle to reach optimal scale [8].

1.2.2.3 Global Energy Politics and Security Concerns

Energetic matters are consistently bound up in geopolitical aspects. The control of petroleum and natural gas deposits drives governments to manipulate their foreign relations, requires military action, and creates unfavorable trade ratios. Energy-exporting countries maintain strong international positions because of their ability to employ resources as diplomatic assets yet import-dependent countries suffer from possible energy-related security risks.

Clean energy transition influences political relationships between world nations. Several Western nations have control over critical minerals needed for batteries and solar panels as well as wind turbines which include lithium, nickel, cobalt, rare earths, leading to supply safety issues and ethics of mining practice and geopolitical advantages.

Renewable energy transformation is redesigning how nations deal with energy affairs. New alliances based on technological advances, supply chain structures, and intellectual property rights formation become possible due to this factor. Countries need to explore various material supplies together with circular economic systems while building essential material stockpiles to reduce their dependence on new energy supplies.

1.2.3 The Structural Challenges of Energy Transition

The energy industry transition to renewable power involves more than the exchange of traditional sources since it demands a complete redesign of production, delivery, and consumption methods. Multiple infrastructure problems need resolution to build a sustainable energy system that delivers resilience for low-carbon operation.

1.2.3.1 Infrastructure Limitations and Grid Adaptation

The existing energy system was built for centralized power stations fueled by fossils rather than renewable generation supported by dispersed units. Electricity grids today have insufficient capability to process power in both directions, adjust consumption depending on demand, or distribute solar and wind energy across space.

Extensive infrastructure transformation of the power system needs to be performed to enable solar power, wind energy, and DERs integration:

- Transmission infrastructure (e.g., high-voltage lines, grid interconnectors)
- Flexibility mechanisms (e.g., demand response, grid-scale storage)
- Real-time control and monitoring systems consist of cyber-physical systems.

Currently existing grid congestion alongside excessive curtailment of renewable energy, and the absence of interregional transmission capacity leads to efficiency drops and revenue reductions. The development of renewable projects faces delays because of permitting issues in combination with land acquisition problems and waiting times for connecting to the power grid.

The transformation of the grid into a modernized system demands investments amounting to multitrillions of dollars along with updates to laws and the cooperation between various authority figures and affected parties.

1.2.3.2 Policy and Regulatory Hurdles

The deployment of clean energy faces significant obstacles from policy doubts coupled with multiple regulatory problems as well as fee structures that are not stable. Resources from renewable energy projects undergo prolonged examination procedures and uncertain national grid requirements in addition to unstable power tariffs that create barriers to funding.

The adoption of carbon pricing mechanisms has grown during the past years, but such systems currently regulate just 23% of worldwide emissions through multiple varying policies. Numerous countries continue to support fossil fuels through subsidies because political and economic influences maintain their position.

The engagement of cross-border trade and scalability remains difficult because of insufficient standardized safety protocols and grid connectivity systems as well as digital energy service specifications. Separate sectors such as energy as well as transport and environment must work together alongside finance and industrial sectors for policy coherence to succeed. Federal institutions need to create long-lasting policies with clear intentions to activate private interests while establishing safety measures for innovative ventures.

1.2.3.3 Societal Resistance and Industrial Pushback

Sustainable energy system transition requires handling social and political effects along with environmental aspects. Local populations tend to oppose renewable energy installations when they believe such projects harm natural environments through wind turbine noise and land disputes and unions oppose fossil fuel phase-outs because personnel will lose their jobs.

Fossil fuel organizations enjoy substantial lobbying influence because they use it to create misleading media messaging that hinders progress on climate policy implementation. Misinformation spread by the public about renewables creates hurdles that impair public backing for their implementation.

A just transition asks for inclusive programs for retraining workers while allowing fair investment distributions throughout society and community-level decision-making to gain consensus from people. When social groups and industries suffering from climate policies protest against them, this could disrupt the nation's climate targets.

1.3 The Acceleration of Renewable Energy Adoption

Global energy policy along with economic development now places renewable energy at the center of its essential frameworks after it started as a marginal environmental preservation movement. The increasing need for carbon reduction leads to an exceptional energy system transformation through technological innovation and cost reductions as well as supportive policy structure enhancements. Renewable energy accelerates through its combination with digitalization and AI which develops reflective, operational, and efficient energy systems.

Nowadays, renewable energy dominates the market primarily because it promotes economic stability as well as strategic positional advantage. Soaring economic development rates in emerging countries lead them to pass over conventional fossil-based infrastructure installations while developed nations establish zero-emission objectives and global financial systems direct their funding toward sustainable technologies. Renewable energy will reach its total capacity after sustained investments into innovative technologies, coordinated policies, and system-wide integrations.

1.3.1 The Core Pillars of Renewable Energy

The renewable energy transformation relies on multiple power sources that bring their specific advantages toward building sustainable and carbon-light energy systems. The key contributors to renewable power generation consist of solar and wind and include hydropower together with geothermal energy and advanced bioenergy solutions for both flexibility and stationary capacity and community energy solutions as seen in Table 1.3 below.

TABLE 1.3

Global Installed Capacity and Economic Metrics for Key Renewable Technologies (2023)

Technology Type	Installed Capacity (GW)	Annual Growth Rate (%)	Estimated Utility-Scale Cost (USD/W)	Relative Maturity	Reference
Solar PV	800	~15	~\$0.70	High	IEA, Lazard's Levelized Cost Studies
Onshore wind	700	~7	~\$1.30	High	IEA, World Energy Council
Offshore wind	75	~20	~\$3.00	Increasing	IEA, Bloomberg NEF
Hydropower	1300	~1	~\$1.20	Mature	IEA, Global Energy Statistical Yearbook
Geothermal	15	~2	~\$2.50	Niche	IRENA, IEA Reports
Bioenergy	100	~3	~\$1.50	Mature/ Evolving	IRENA, World Bioenergy Report

1.3.1.1 Solar and Wind Power: The Driving Forces

Solar photovoltaic (PV) systems together with onshore/offshore wind energy have become the leading renewable energy choices because of their adaptive capabilities, price decreases, and their natural fit for distributed power generation. Solar and wind energy systems have experienced a reduction in Levelized Cost of Electricity by more than 80% since the beginning of the previous decade which has made them less expensive than fossil-fuel-based power generation methods [9].

Solar PV development with floating and bifacial solar module deployment having perovskite material utilization could expand commercial solar generation opportunities as we could see in Figure 1.2 that how fast is the growth of harvesting energy from PV in near future as compared to other renewable resources. Wind energy production achieves optimal performance from design improvements of turbines, better capacity utilization, and digital twin applications. The expansion of offshore wind energy relies heavily on floating platforms because these installations open high-wind areas to power generation. The integration problems related to intermittency find solutions through hybrid systems and storage units, and AI-based forecasting which help make wind and solar power more dependable for contemporary power networks.

1.3.1.2 Hydropower, Geothermal, and Next-Gen Bioenergy

Hydropower is currently the dominant renewable power generation method worldwide because it gives stability to power grids and offers both dispatch capability and storage possibilities through pumped storage technology.

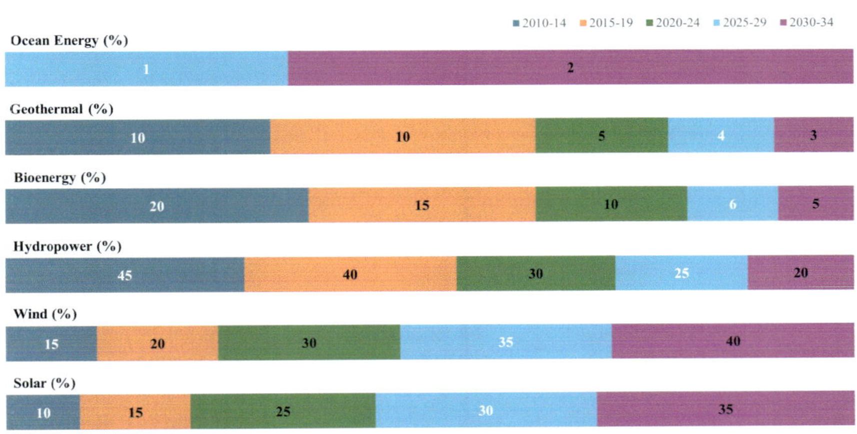

FIGURE 1.2
Renewable energy mix evolution (2010–2030). (*Source*: BP energy review, 2023.)

Scarce funding is currently directed toward developing small micro-hydro systems because they offer localized electricity generation for off-grid areas.

Enhanced geothermal systems have attracted attention because they extract resources from areas that were formerly inaccessible for geothermal power generation.

Bioenergy development progresses via second-generation biofuels together with biomass gasification processes which minimize land use wars while totaling carbon emissions to zero. The third generation of bioenergy makes use of algae along with waste-to-energy solutions and synthetic biology for creating specific energy applications. The combination of these sources provides intermittent support for renewable supplies which aids the operation of heavy industry and transportation sectors.

1.3.2 The Role of AI and Digitalization in Renewable Energy

Renewable energy receives a transformation through AI and digital technologies which provide three key capabilities, namely predictive analytics, autonomous control, and system-wide optimization. Table 1.4 describes below the transportation of energy through AI, which results in higher operational

TABLE 1.4
AI Integration Across the Renewable Energy Value Chain

Stage in Energy Value Chain	AI Applications	Expected Benefits	Key Challenges	Reference
Generation	Predictive maintenance, solar/wind forecasting	~10–20% improvement in capacity factor, reduced downtime	Data quality and sensor integration	IEA Technology Roadmaps
Transmission	Dynamic load balancing, fault detection	Enhanced grid stability; reduced line losses by up to 5–10%	Cybersecurity risks	Smart Grid Reports, IEA
Storage	Battery management optimization, real-time charge/discharge scheduling	Extends battery life by ~15%, cost efficiency improvements	Modeling of degradation parameters	Journal of Energy Storage, IEA
Distribution	Demand-response systems, smart metering	~10–15% demand reduction during peak times, improved grid resilience	Consumer adoption, infrastructure upgrades	Smart Grid Reports, IEA
Consumption	Energy management systems, AI-based home automation	Improved energy efficiency in households (up to 20% savings)	Integration with legacy systems	Research in Energy, IEA Reports

effectiveness and better power reliability as well as increased integration of fluctuating energy resources into network systems.

1.3.2.1 AI in Energy Demand Forecasting and Grid Optimization

The accurate prediction of future demand represents an essential aspect for balanced supply and demand operations in renewable energy power systems. AI models that process information through ML and NN, and big data provide us with better insights into energy demand forecasting. The system gathers information through the combination of weather data evaluation along with user activities and market behavior insights.

The models support the following functions: dynamic balance for grids along with lowered curtailment methods and real-time operation for dispatch and load control. Predictive maintenance tools operated by AI systems assist in increasing renewable energy system reliability by detecting problems ahead of time, thus extending the lifetime and maintaining system operational steadiness.

As a result of AI-driven power grid technology, voltage control, load balancing, and congestion control have all been enhanced, which is clearly visible in Figure 1.3 that in the coming era, how much it is going to be impactful [10]. The entire energy system functions with better efficiency and operational resilience because of this development specifically when multiple power sources become prevalent.

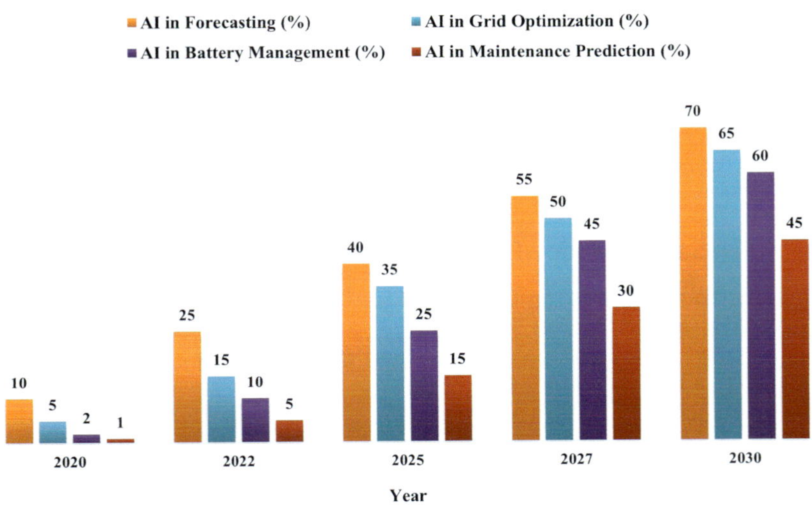

FIGURE 1.3
AI integration in renewable energy systems (2020–2030). (*Source*: IRENA AI report, 2022.)

1.3.2.2 Smart Grids and Decentralized Energy Networks

AI together with IoT develops adaptive energy systems that operate in two directions to control DERs extending from rooftop solar to EVs and home batteries. Smart grids achieve their features through the combination of edge computing with intelligent sensors.

- Autonomous demand response
- Peer-to-peer energy trading
- Microgrid formation during outages

AI systems work to instantly modify power system structures for automatic fault detection to reduce time without an electricity supply and boost the network's defense capabilities. The mentioned features prove highly beneficial for regions facing risks from both natural disasters and geopolitical instabilities.

1.3.2.3 AI-Enhanced Battery Storage and Energy Efficiency

BESS represents an essential element that helps manage solar and wind power fluctuations. AI enhances storage performance through:

- State-of-charge estimation
- Predictive degradation models
- Dynamic charge-discharge scheduling

The implementation of these optimizations allows batteries to last longer while decreasing operation expenses and achieving short response times in grid systems. It also supports intelligent energy monitoring, load prediction, and behavior analysis. It impacts on energy efficiency for both industry and homes.

1.3.3 Financial and Policy Enablers of Renewable Growth

Renewable energy development must have financial tools together with planning methodologies and worldwide partnerships for extended expansion to occur. A positive investment environment together with risk reduction leads to accelerated innovation.

1.3.3.1 Government Incentives and Carbon Pricing

Many factors exist in public policy, yet none holds greater influence than its ability to construct renewable energy markets. Renewable energy adoption receives strong support from FiTs, PTCs, ITCs, and renewable portfolio standard (RPS), which have demonstrated effectiveness in promoting renewable adoption. The market has adopted auction-based pricing together with green bonds along with performance-based incentives as new financing methods.

The implementation of carbon taxes alongside cap-and-trade systems through proper internalization of environmental costs allows renewables to gain higher competitiveness. Such mechanisms and structures drive the creation of advanced carbon capture solutions and emerging low-emission technology innovations. The key elements of financial incentive include policy consistency coupled with clear disclosure and sustained planning, which promote both investing capital and infrastructure establishment.

1.3.3.2 Private Sector Investments and Market Disruptions

The expansion of renewables is done by way of venture capital together with corporate power purchase agreements and green financial investments in the private sector. Significant investments from technology businesses along with energy corporations occur as they pursue ESG objectives while simultaneously working to decrease operational challenges.

The energy sector embraces new market disruption through the energy-as-a-service (EaaS) model and blockchain trading with virtual power plants (VPP) technology which transforms power economy dynamics. The new technological developments make clean energy more accessible to all while erasing entry points for businesses, especially in areas without sufficient services.

Institutional investors continue to sell their fossil investments because they support the worldwide momentum toward sustainable financial assets and clean technological alternatives.

1.3.3.3 Global Collaborations and Climate Agreements

The deployment of renewable energy networks requires international consensus along with solutions for addressing international climate change issues. The Paris Agreement alongside the International Renewable Energy Agency (IRENA) facilitates knowledge exchange and transfers technology systems as well as developing unified policy guidelines.

The sustainability objectives of public and private stakeholders unite through collaborative programs including Mission Innovation, Clean Energy Ministerial, and Powering Past Coal Alliance. Through these platforms progress moves faster because they provide R&D financing in addition to standardization support and energy access initiatives for developing nations.

Several cross-border power systems, international research programs, and joint investment ventures are leading to a worldwide interconnected sustainable energy system.

1.4 Opportunities and Challenges in Scaling Renewables

The worldwide effort to transition toward renewable energy systems enables major transformational prospects but requires handling multiple complexities. Countries recognize that the time has arrived to implement renewable energy deployment rather than debating its feasibility for achieving decarbonization and climate goals. This segment discusses macro-scale advantages of renewable energy economies in addition to the necessary framework changes for justifying and stabilizing the energy transition process.

Several scientific disciplines must unite for successful national and international renewable integration, and this alliance includes technological development along with strategic policy structure, power networks, and societal support. The fast improvement in renewable technologies does not resolve the key barriers preventing the sustainable scalability of renewable energy systems.

1.4.1 Benefits of a Renewable Energy Economy

Swapping to renewable energy systems results in various advantages that transcend environmental preservation. Energy systems today adopt renewable energy fundamentals which create new ways national economies contribute to stability and enduring economic success.

1.4.1.1 Reducing Carbon Emissions and Climate Mitigation

Renewable energy is essential in achieving all objectives under the Paris Agreement's climate targets. The GHG emissions output from solar, wind, hydro, and geothermal power remains minimal. Major renewable energy adoption reduces global carbon emissions most strongly within sectors of power generation, industry, and transport because they emit the most carbon.

Utilities powered by renewable sources provide the opportunity to cut emissions throughout sectors that transition toward electric systems (for instance, electric vehicles and heat systems). Through comprehensive life cycle assessments (LCAs) of renewable technologies, producers establish beneficial emission results which take into consideration all stages from manufacturing through installation up to decommissioning.

1.4.1.2 Energy Independence and National Security

National energy independence grows through renewables which reduces the vulnerability of shifting between unpredictable fossil fuel market prices and imports. Targeting solar energy, wind resources, and biomass residues within local boundaries enables countries to produce their electricity and shield themselves against international energy problems such as interruptions, price instability, and geopolitical conflicts.

Decentralized renewable power systems establish better resilience through independent local energy distribution when grids lack functionality because of both terrorism and natural disasters.

1.4.1.3 Job Creation, Economic Growth, and Social Impact

Green jobs receive their main infrastructure support from the renewable energy sector which operates through various stages including manufacturing, installation, operations, and research and development. Statistical data from IRENA demonstrates that renewable energy generated 13 million job opportunities throughout the world during 2023 and expects to add more employment in the coming years.

The sector encourages regional economic growth through off-grid solar and micro-hydro systems that help establishments in remote areas to deliver better services to local communities. Where energy access programs are designed for inclusiveness, they provide benefits to deprived communities while lowering energy poverty levels and creating social equality.

1.4.2 Overcoming the Key Challenges

The renewable energy transition encounters two types of barriers which need specific interventions due to their technical nature alongside systemic obstacles. The widespread adoption of renewable energy faces resistance from storage restrictions along with insufficient basic infrastructure and long obstacles in social procedures.

1.4.2.1 Energy Storage and Grid Stability Solutions

Solar and wind energy systems generate power in ways that create problems for established power supply-demand management methods in electrical grids. Grids become unstable along with higher curtailment risks when storage capacities and quick response demand do not exist adequately.

Different advanced storage solutions such as lithium-ion batteries along with flow batteries, solid-state batteries, and emerging thermal energy and hydrogen storage capabilities serve as major assets to smooth power interruptions while permitting time-based energy distribution. AI-controlled battery management systems operate more efficiently while managing the life patterns of storage systems.

The power grid becomes more flexible because demand-side management techniques and flexible power-generation systems (hydro or gas peakers) work in combination with inter-regional electricity transfers. The operation of high-renewables power systems requires immediate dynamic frequency adjustments together with immediate voltage control and natural inertia support to maintain stability.

1.4.2.2 Transitioning from Centralized to Decentralized Energy Models

Large-scale energy generation exists in central places before power flows unidirectionally to passive distribution networks. The renewable energy generation model transforms power systems because it allows different-sized power producers including residential PV installations, community wind facilities, and industrial rooftop systems.

Energizing decentralized systems demands a complete grid system redesign along with fresh regulations and market systems. The three enabling technologies for energy democratization include intelligent meters alongside data analysis systems and peer-to-peer trading processes that use blockchain technology.

Current difficulties within power management systems and tariff structures remain unacceptable throughout non-developed economies since they block equal infrastructure improvements.

1.4.2.3 Addressing Political, Institutional, and Social Barriers

The process of expanding renewable energy systems faces similar barriers of governance management as it does technical issues. Policies create uncertainty and bureaucratic systems along with fossil fuel industrial pressure and improper reward structures tend to delay renewable energy projects.

The process of renewable development faces hurdles when institutions delay project authorizations or show poor performance in establishing new standards or connecting to power grids. Social acceptance depends on how people perceive the visual implications of renewable systems and also involves disputes regarding land management and incorrect information spread throughout the public.

A transition strategy needs multiple elements that combine stakeholder involvement with planning participation and financial support for displaced fossil fuel workers and public education training for building social acceptance. Effective policy alignment between different international territories together with border-spanning collaborations serves as a fundamental requirement to hasten deployment while adhering to climate-related agreements.

1.5 Future Outlook: The Path to a Sustainable Energy Ecosystem

The path to energy sustainability passes through three essential elements of smart technology, innovation, and environmental preservation. The urgent need to evolve energy infrastructure becomes crucial because climate deadlines pressure an environment of increasing worldwide energy consumption. The subsequent chapter puts together technological, political, and scientific innovations which will unite to create resilient low-carbon energy systems along with intelligent energy networks.

The upcoming stage of energy transformation depends on a quick interface between power generation and storage systems with distribution networks along with user consumption methods. The course of energy system development which leads to sustainable growth depends on AI together with international policy frameworks and frontier research.

1.5.1 The Future of AI-Driven Energy Systems

Future energy infrastructures will receive their nervous system designation from AI systems. Energy systems can build real-time data and ML algorithms for dynamic control, optimization, and prediction throughout the entire energy supply chain.

Machine learning enhances the performance of grid operations when used for load balancing, fault identification, and asset maintenance as well as cybersecurity measures. Through its functionality, it enhances energy storage efficiency by managing battery cycles while it both extends battery systems' operational time and anticipates system deterioration.

AI functions to synchronize the distribution of electricity with the current grid capacity in real-time operations. The entire system operates more efficiently through this process while preventing high-energy use periods. VPPs along with microgrids have broad advantages from AI technology because it facilitates independent management combined with market intelligence.

The development of dynamic pricing systems and auction-based renewables entry alongside peer-to-peer energy transactions happens because of AI. Future power systems built for high renewable integration will need AI systems as their core requirement because these systems become more challenging to manage due to multiple distributed intermittent sources.

1.5.2 Global Policy Roadmaps and Energy Transition Scenarios

The achievement of sustainable energy needs governments to unite their policy strategies under transition plans that support global climate targets. Various governments across the world develop prolonged decarbonization approaches by creating Nationally Determined Contributions (NDCs) together with net-zero pledges and defining sector-specific emission limits.

Many organizations including the IEA together with IPCC and national planning bodies develop structured predictions of future energy system evolution through multiple policy and behavioral and technological scenarios. The analysis examines multiple factors involving carbon pricing along with electrification speeds and technology affordability together with changes in human conduct.

Moving forward, the main policy measures will involve:

- Environmental expenses can be integrated through carbon market platforms which function as emissions trading systems.
- Recovery technologies receive support through two clean energy programs called RPS and feed-in tariffs.
- Green industrial policies for local manufacturing, workforce training, and R&D funding.
- The energy transition should implement justice frameworks that promote equal access for everyone to participate along with fair distribution throughout the system.

National efforts for sustainable development must be strategically connected to worldwide objectives which depend on the United Nations Framework Convention on Climate Change (UNFCCC), Clean Energy Ministerial together with Mission Innovation to achieve this goal [11].

1.5.3 The Role of Innovation and Research in a Green Economy

A green economy relies on constant innovation mechanisms together with research work across various disciplines. Future advancements should unite technological progress and system-wide conjunction efforts because these elements work together to overcome present constraints regarding system capabilities and cost-effectiveness, and expand operational boundaries.

Key research frontiers include:

- Next-generation energy storage (e.g., solid-state batteries, metal-air, and thermal storage).
- Emerging renewable technologies such as floating solar, advanced wind turbines, and ocean energy.
- Power-to-X solutions (such as green hydrogen as well as synthetic fuels) bridge electricity production with challenging-to-abate industrial sectors.
- Heavy industries can use carbon capture utilization storage (CCUS) as a temporary decarbonization support method.

The field of materials science contributes significantly to improving efficiency along with extending service times and developing recycling capabilities of renewable technologies. The combination of quantum computing with AI technology has sped up the identification of new battery substances and PV components.

Irrational efforts between the public and private sectors together with innovation centers and open research networks will help sustainably upscale green technologies. In order to achieve global deployment, it will be important to develop local innovations that suit specific resources together with climate conditions and socio-economic situations.

A sustainable global energy system built on fair principles can be created by building an innovative ecosystem which includes academic along with industrial components and cooperation between policy-making authorities and civil groups.

1.6 Conclusion: Building the Bridge to a Sustainable Energy Future

The world now requires an energy systems transition toward clean, intelligent, and resilient energy ecosystem that prioritizes both greater energy supply and climate protection because the situation has evolved beyond being optional. The base knowledge for comprehending critical global energy problems and the process to adopt renewable power while developing robust storage networks and applying AI-enhanced technologies became established through this chapter.

Our analysis defined the ecological hazards and financial challenges as well as geopolitical consequences stemming from the fossil fuel-based system while clarifying the blocking elements that originate from both procedural and public and business obstacles. Renewable energy deployment receives worldwide support because modern technology and strategic policy structures together with AI enhance system optimization.

The future of energy receives its transformation through renewables which include solar power, wind power, hydropower, and developing bioenergy solutions. However, their intermittent nature necessitates a parallel transformation in energy storage and smart grid technologies. AI functions as a critical element in predicting customer needs as well as operating distributed energy systems and boosts system performance.

Sustainable energy systems will achieve their maximum scalability and success by implementing visionary policies and global cooperation between researchers and developing inclusive implementation strategies. A comprehensive collaboration between governments together with academia, industry, and civil society organizations will maintain both fairness and sustainability of energy transition benefits.

This introductory chapter establishes both the energy crisis urgency and renewable system growth opportunities which guide the academic investigation into AI and renewable energy synergy practices of this publication. Advanced analytics capabilities coupled with operational intelligence and new energy architectures will set the path toward a zero-emissions or decentralized energy future and data monitoring.

References

1. https://www.iea.org/reports/world-energy-outlook-2023/executive-summary
2. https://www.thesunprogram.com/research-platform-feed/ipcc-intergovernmental-panel-on-climate-change?gad_source=1&gad_campaignid=22228028095&gbraid=0AAAAACU23rPPFviW6ZMCC7cH7wxE_P7fT&gclid=CjwKCAiA09jKBhB9EiwAgB8l-BgjnLlOCmc5g2ghI4WL4JyqpPaOW21At6AA7m4W6Q0c9VYfn0ZgdxoCDMMQAvD_BwE
3. https://www.iea.org/reports/sdg7-data-and-projections/energy-intensity
4. https://www.irena.org/Digital-Report/Renewable-energy-and-jobs-Annual-review-2023
5. https://sdgs.un.org/goals
6. https://www.iea.org/reports/co2-emissions-in-2022
7. who.int/health-topics/air-pollution
8. https://sdgs.un.org/goals/goal7
9. https://www.bp.com/content/dam/bp/business-sites/en/global/corporate/pdfs/energy-economics/energy-outlook/bp-energy-outlook-2023.pdf
10. https://www.irena.org/Digital-Report/World-Energy-Transitions-Outlook-2022
11. https://unfccc.int/process-and-meetings/united-nations-framework-convention-on-climate-change

2

Renewable Energy Technologies and Energy Storage Systems

Yasir Yasin and Veerta Mah-i-Liqua

2.1 Introduction

Global energy demand is escalating due to industrialization, urbanization, and population growth. The reliance on fossil fuels has resulted in significant environmental degradation and climate change. Renewable energy technologies (RETs) provide sustainable options by harnessing natural resources such as solar, wind, biomass, geothermal, and hydropower. However, the intermittent and fluctuating characteristics of numerous renewable sources, particularly solar and wind, constrain their dependability and potential for grid integration. Renewable Energy Storage (RES) systems are essential for providing a stable, sustainable, and more efficient energy supply to overcome these challenges.

Renewable Energy Storage (RES) technologies collect surplus renewable energy during periods of low demand or elevated generation and release it when needed, thereby enhancing energy utilization, system stability, and overall efficiency. Some of the renewable energy storage systems are Concentrated Solar Power (CSP), photovoltaic-thermal (PV/T) systems, and biomass-based combined heat and power (CHP) plants, etc.

2.2 Renewable Energy Technologies (RETs)

Depletion of fossil fuels, an increasing trend of greenhouse gas (GHG) emissions, and the mitigation of climate change are the responsible factors for the swift change of the global energy landscape. The following renewable energy technologies are currently being used for energy generation worldwide,

1. Solar Energy
2. Wind Energy

DOI: 10.1201/9781003516132-2

3. Chemical Energy – Biomass and Biochemical Energy
4. Geothermal Energy

These energies not only resolved the problem of sustainable energy resources but also showed a promising low environmental impact. The combination of these technologies with effective energy storage systems (ESS) is crucial for addressing intermittency and attaining high levels of integration in power grids. A descriptive note for all the above-mentioned energies will later be discussed in this chapter.

2.2.1 Principles of Renewable Energy Conversion

The renewable energy conversion principles are the fundamental thermodynamic laws and principles. As per the first law of thermodynamics (energy conservation),

$$\Delta U = Q - W$$

where,
$\Delta U = Change\ in\ Internal\ Energy,\ Q = Heat,\ W = Work$

The direction of energy flow is governed by the second law of thermodynamics. As per the second law of thermodynamics, the overall efficiency of a renewable energy system is given by the following relation,

$$\eta = \frac{W_{useful}}{Q_{input}}$$

where,
$\eta = Overall\ Efficiency\ of\ Renewable\ Energy\ System$
$W_{useful} = Amount\ of\ Useful\ Work\ Produced$
$Q_{input} = Heat\ Input\ to\ the\ system$

The conversion mechanisms are different for different renewable energy systems, such as

Solar Energy – Solar radiation converted into electricity or thermal energy

Wind Energy – Kinetic energy of wind is converted into mechanical energy by wind turbines

Chemical Energy – Chemical energy in biomass is utilized through combustion, gasification, or digestion

Geothermal Energy – Uses the subsurface energy for heating or power generation purposes

The exergy analysis not only evaluates the performance of renewable energy systems but also identifies the areas of improvement. The following relation is used for the evaluation of the exergy of the system,

$$\psi = \frac{W_{useful}}{W_{rev}}$$

where,

$\psi = Exegy\ of\ the\ system$

$W_{useful} = Maximum\ Amount\ of\ Useful\ Work\ Produced$

$W_{rev} = Maximum\ Reversible\ Work\ that\ can\ be\ produced$

Some of the renewable energy technologies will be discussed hereafter.

2.2.2 Solar Energy Technologies

The abundant energy flux from the sun, which is estimated to be 17×10^{17} W incident on Earth's atmosphere, is captured with the help of solar energy technologies. Photovoltaic (PV) and concentrated solar power (CSP) systems are two ways to use solar energy, and each has unique thermodynamic and engineering properties.

2.2.2.1 Photovoltaic (PV) Systems

PV systems use semiconductor materials that use the photovoltaic effect for the conversion of solar radiation into electrical energy. A schematic diagram is shown in Figure 2.1. The following relationship is used to evaluate the performance of PV systems [1]:

$$I = I_{ph} - I_o\left(e^{\frac{V}{nkT}} - 1\right)$$

where,

$I_{ph} = Photogenerated\ Current,\ I_o = Reverse\ Saturation\ Current$

$q = Electron\ Charge,\ n = Ideality\ \ factor$

$k = Boltzmann's\ Constant,\ T = Temperature$

The maximum power output occurs at the maximum power point (MPP), characterized by the efficiency,

$$\eta_{PV} = \frac{P_{max}}{GA}$$

where,

$\eta_{PV} = Efficiecy\ of\ \ PV,\ P_{max} = Maximum\ Power\ Output,$

$G = Solar\ Irradiance,\ A = Cell\ Area$

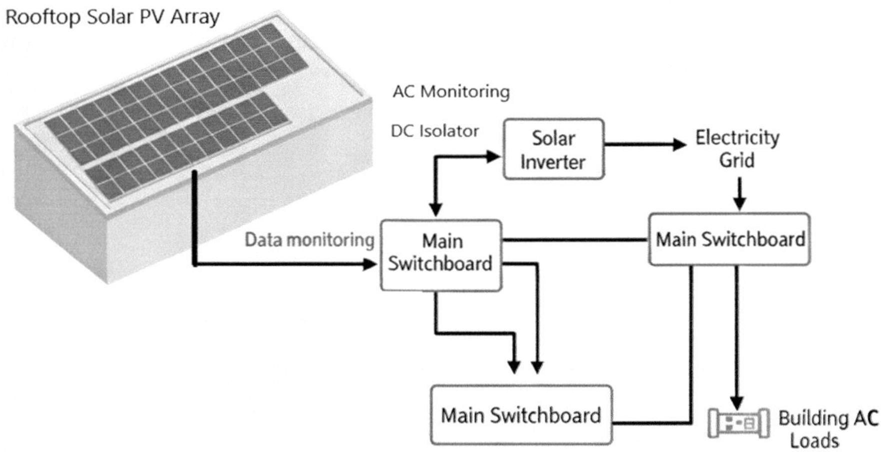

FIGURE 2.1
Schematic diagram of solar PV system. (*Source*: Author.)

The efficiency and voltage generation are greatly affected by the module temperature. The heat transfer analysis is used to forecast module temperature. Therefore, the system efficiency is enhanced by effectively transferring the temperature and integrating maximum power point tracking (MPPT) systems [2]. Moreover, Perovskite–silicon tandem cells, bifacial modules, and multi-junction cells (which surpass 45% efficiency) are examples of engineering advances in photovoltaics.

2.2.2.2 Concentrated Solar Power (CSP) System

Parabolic troughs, linear Fresnel reflectors, power towers, and parabolic dishes are examples of optical concentrators used in CSP technologies to direct sunlight onto a receiver that heats a heat-transfer fluid (HTF). Then, this thermal energy is used to produce electricity. The rate of useful thermal energy is given by the following relation [3],

$$\dot{Q} = \dot{m} C_P \left(T_{out} - T_{in} \right)$$

And the overall optical thermal efficiency of the system can be given as

$$\eta_{CSP} = \eta_{opt} \eta_{th} \eta_e$$

where

$\eta_{opt} = Optical\ Efficiency$, $\eta_{th} = Thermal\ Conversion\ Efficiency$
$\eta_e = Turbine\ and\ Generator\ Efficiency$

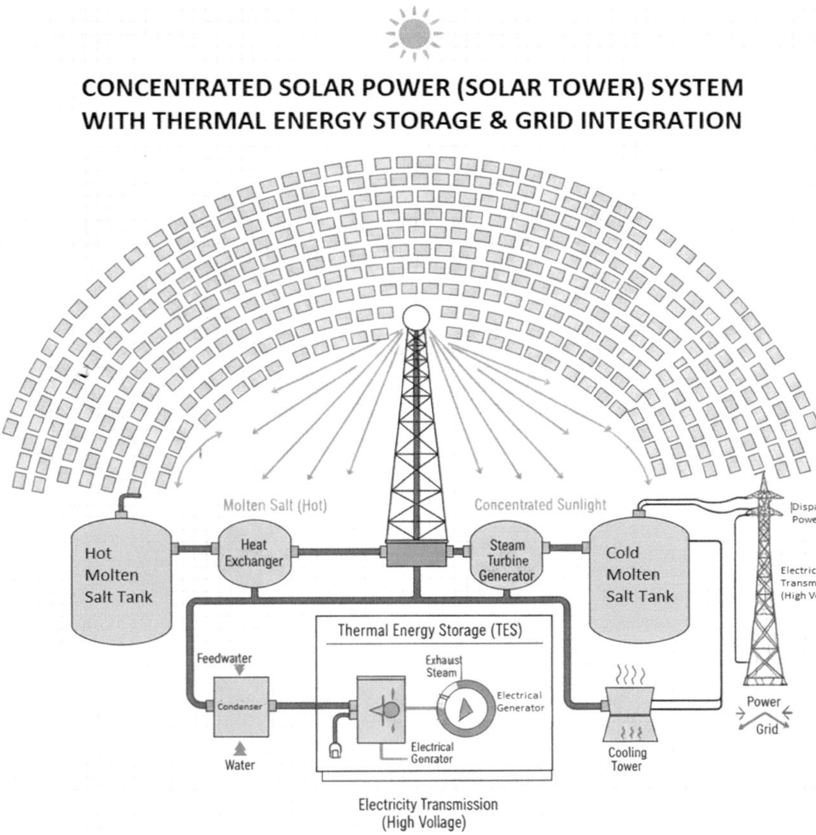

FIGURE 2.2
Schematic diagram of CSP (Solar tower) system. (*Source*: Author.)

Figure 2.2 provides a schematic diagram of a concentrated solar thermal power system integrated with the grid for onward distribution.

2.2.2.3 Summary of Solar Energy Technologies

Table 2.1 represents a brief summary of different solar energy technologies emphasizing their working principles, electrical efficiencies, temperature range, advantages, and limitations.

2.2.3 Wind Energy Systems

The basic principle of the wind energy system is to convert the kinetic energy of the moving air to mechanical energy and then to electrical energy with the

TABLE 2.1

Summary of Different Solar Energy Technologies

Solar Energy Technology	Working Principle	Electrical Efficiency (%)	Temp. Range (°C)	Advantages	Limitations
Photovoltaic (PV)	Conversion through Semiconductors	15–22	25–80	Low Capital & Maintenance Cost, Direct Electricity Generation, Modular System, Easy Capacity Enhancement	Intermittent Source of Energy, Low Efficiency, Performance Affected by Temperature Changes
Parabolic Trough	Concentrated Solar Thermal Power (CSP)	14–25	150–400	Compatible Integration with TES, Moderate to High Temperature Range, Scalable System	Single-Axis Tracking Limitation, High Capital Cost, Thermal and Optical Losses
Solar Tower	Single Point Focus of Solar Thermal Energy	16–32	500–1200	Compatible with Large Scale TES Integration, Very High Temperature Range, High Efficiency	High Capital Cost, Challenging Maintenance
Linear Fresnel	Line Focus of Solar Thermal Energy	14–20	200–450	Simple, Compact & Low-Cost System, Easy Maintenance	Low Efficiency, Low Temperature Ranges
Dish Stirling	Single Point Focus to Operate Stirling Engine	30–35	600–1200	High Efficiency, Modular System, Water Free System	Complex and Expensive System, TES Integration Complexity

help of a generator. The following relation is used to evaluate the available power from wind,

$$P_{wind} = \frac{1}{2}\rho A v^3$$

where

$\rho = Air\ density,\ A = Rotor\ Swept\ Area\ and\ v = Wind\ Velocity$

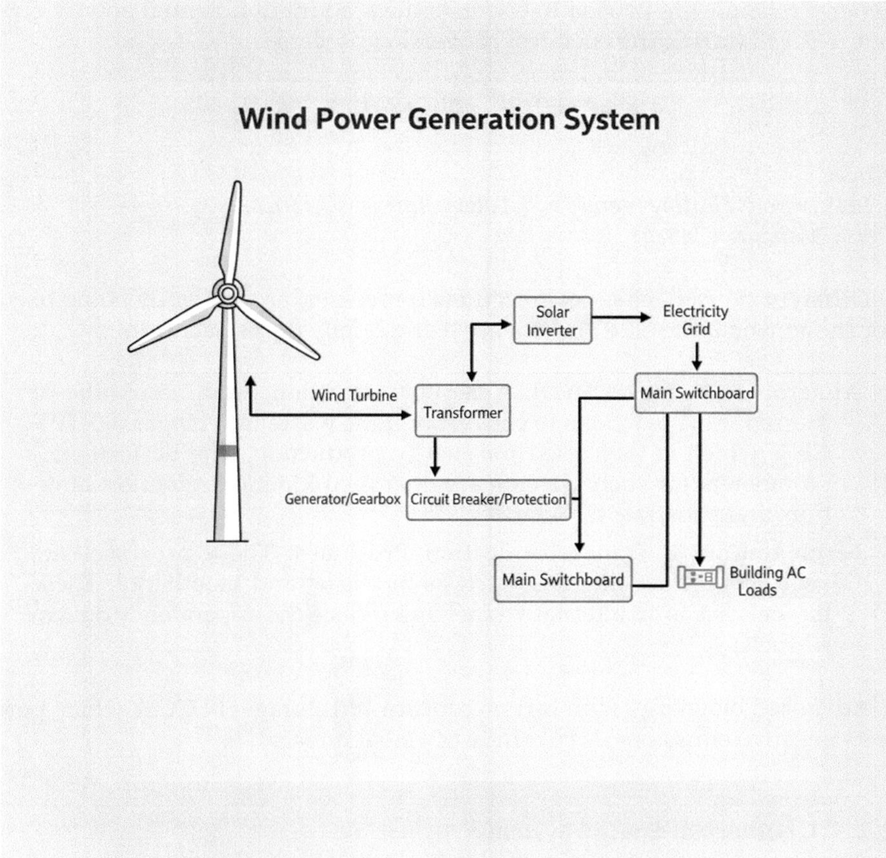

FIGURE 2.3
Schematic diagram of wind power system. (*Source*: Author.)

A schematic diagram of a wind power system is shown in Figure 2.3. Moreover, for variable-speed operation, modern wind turbine design incorporates power-electronic converters, composite blade materials, and aerodynamic optimization.

2.2.4 Chemical Energy Systems

Chemical processes are used to convert the organic material into alternative fuels and utilize the heat produced in these reactions for heating of heat transfer fluids (HTFs). Anaerobic digestion, gasification, pyrolysis, and combustion are examples of thermodynamic conversion methods. The following general relationship is used to estimate the amount of potential energy output released during the chemical conversion process,

$$HHV = LHV + h_v(w)$$

where
$LHV = Low\ Heating\ Value,\ h_v = Latent\ heat\ of\ Vaporization$
$w = Moisture\ Content$

Different types of chemical reactions (processes) are currently being used for the conversion. Some of them are being briefly discussed as follows,

Anaerobic Digestion (AD): Anaerobic digestion uses mesophilic or thermophilic reactions to convert organic waste into biogas (60–70% CH_4), which is then used for energy production. The performance parameters for such systems are microbial kinetics, retention duration, and substrate C/N ratio.

Fermentation & Transesterification Processes: These processes are used to produce biofuels, such as biodiesel and bioethanol. These biofuels act as an alternative fuel and reduce the dependency on fossil fuels.

Advanced bioenergy with carbon capture and storage (BECCS), which permits negative emissions, is currently of global interest [4].

2.2.5 Geothermal Energy Systems

In these systems, geothermal energy is being converted into different energy-producing cycles. Geothermal energy uses crustal gradients and subsurface heat from radioactive decay, and the global heat flux is approximately 4.5×10^{13} W [5]. The power output by a geothermal well is given by the following relationship,

$$P = \dot{m}\left(h_{prod} - h_{inj}\right)\eta_{th}$$

where
$\dot{m} = Mass\ flow\ rate,\ w = Moisture\ Content$
$h = Specific\ enthalpy\ of\ produced\ and\ reinjected\ fluid$

Dry steam, flash steam, and binary cycle plants are the three types of geothermal systems. The latter uses the Kalina Cycle or Organic Rankine Cycle (ORC), which works well with low-enthalpy resources (less than 150°C). Figure 2.4 shows a schematic diagram of geothermal energy system.

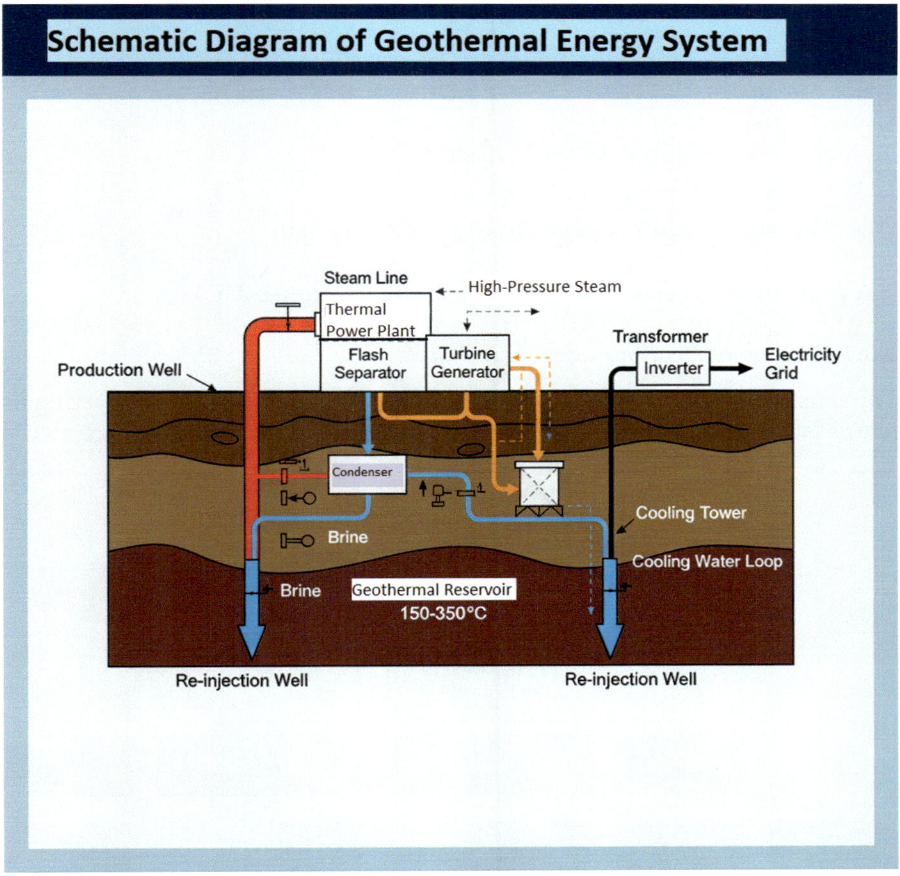

FIGURE 2.4
Schematic diagram of geothermal energy system. (*Source*: Author.)

A combination of heat pumps and CSP systems with geothermal energy is used to create hybrid systems that provide renewable energy uninterruptedly. These hybrid systems are currently being used worldwide for power generation as well as heating applications.

2.3 Thermal Energy Storage Systems

This section will cover a brief introduction to the renewable energy storage systems and their integration with the renewable energy technologies (RETs).

2.3.1 Classification of TES Systems

Thermal Energy Storage systems can be classified into the following three major categories,

1. Sensible Heat Storage (SHS) System
2. Latent Heat Storage (LHS) System
3. Thermochemical Energy Storage (TECS) System

A more detailed classification is illustrated in Figure 2.5.

2.3.1.1 Sensible Heat Storage (SHS) System

The sensible energy is being stored in the heat storage medium by elevating its temperature and keeping its phase the same. Different energy storage media

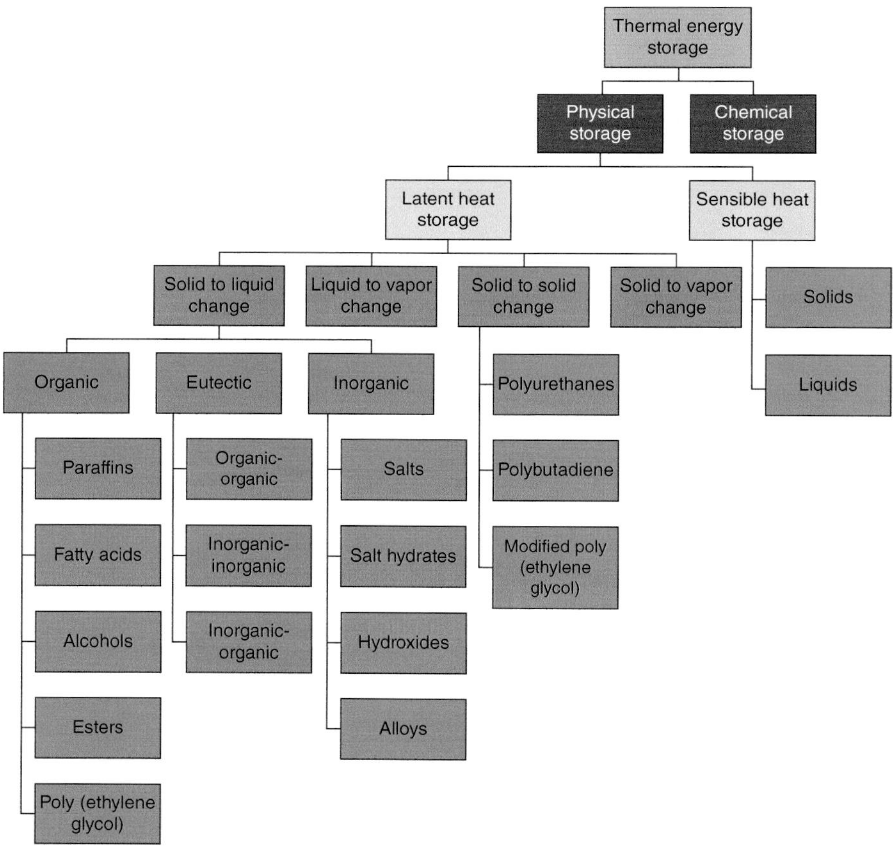

FIGURE 2.5
Classification of thermal energy storage systems. (From [6].)

are currently being used for this purpose, such as water, rocks, molten salts, etc. Among these materials, molten salts are commonly used in CSP systems, and water is being used as a source of heat storage in low-temperature solar thermal systems. The molten salts are based on Na-K-based compounds; the details related to molten salts will be discussed in the later chapters of this book as well. The following relation is used for estimating the stored energy in different media [7].

$$Q = m\ C_p\ \Delta T$$

where,

$m = mass\ flow\ rate\ of\ material,\ \Delta T\ = Temperature\ difference.$
$Cp\ = Specific\ heat\ at\ constant\ pressure$

These systems are simple, low cost, and have a wide temperature range. Their limitations are a large volume requirement, low energy density, thermal losses, and non-isothermal processes involved.

2.3.1.2 Latent Heat Storage (LHS) System

The latent heat (energy) is stored in the heat storage medium by changing its phase and keeping its temperature constant. The energy is absorbed during the charging time (heat storage time), and the same amount of energy is released during the discharging time. Different phase change materials (PCMs) are used for this purpose, such as paraffin wax, erythritol, salts, etc. These materials have high latent heat of fusion and are perfect for solar heating and process heat applications because they maintain a nearly constant temperature during melting and solidification.

The recent development in these systems includes encapsulated PCMs and nanocomposite PCMs with improved conductivity [8].

The advantages of these systems include high energy density, compact design, and isothermal processes. Their limitations include limited stability, low thermal conductivity, and higher cost.

2.3.1.3 Thermochemical Energy Storage (TCES) System

Reversible chemical reactions are used for the storage of energy in these storage systems. Different types of chemical compounds are used for this purpose, such as metal oxides, carbonate systems, etc. To store and release energy, thermochemical systems use reversible reactions like metal oxide reduction-oxidation reactions (e.g., $CaCO_3 \leftrightarrow CaO + CO_2$) [9]. Despite their high energy density and potential for long-term storage, these systems struggle with material degradation and reaction kinetics.

The advantages of these systems include high-temperature applications, the highest energy density, and minimal storage losses, and the limitations are high capital cost, complexity of design, and material degradation.

2.4 Integration of TES with Renewable Energy Systems

The intermittency of renewable energy technologies is resolved with the integration of energy storage systems. Different types of energy storage systems are used for different renewable energy technologies; some of these are being discussed hereafter.

2.4.1 Integration of Energy Storage System with Solar Energy Systems

The integration of an energy storage system with solar energy systems increases the system reliability and sustainability by providing power even in the absence of sunlight. The energy storage system is selected on the basis of the renewable energy technology being used. Photovoltaic (PV) systems are being integrated with the battery packages to provide energy during off-hours. Similarly, concentrated solar power (CSP) systems are integrated with different thermal energy storage systems, such as paraffin wax, pebbles, or molten salt tanks, etc. Both sensible as well as latent heat storage media can be integrated with CSP systems; the selection is based on the temperature ranges provided by the CSP systems. An example of this integration is the Gemasolar plant in Spain, in which a 15-hour molten salt storage period with a capacity factor of greater than 75% is demonstrated [10]. Therefore, PV-T and TES systems improve overall energy use by combining thermal and electrical outputs. Figure 2.6 is the schematic diagram showing integration of TES with of a solar PV system. It also represents the power generation, its transmission, and storage.

2.4.2 Integration of Energy Storage System with Wind Energy

Wind's intermittent nature makes it necessary to integrate energy storage systems (ESS). This provides a more reliable source of energy and reduces the intermittency of this renewable energy technology. Compressed air energy storage (CAES) systems and hybrid systems are used for this purpose. These systems couple an energy storage system with wind power generation systems. The air is compressed when there is sufficient wind velocity, and this compressed air's energy is then utilized to provide the power during the off-hours by expanding it. By recovering waste heat from compression and using it during expansion, TES integration in wind-based systems improves overall cycle efficiency, increasing round-trip efficiency from 40% to almost 70% [11]. Figure 2.7 shows the integration of an energy storage system with a wind energy system.

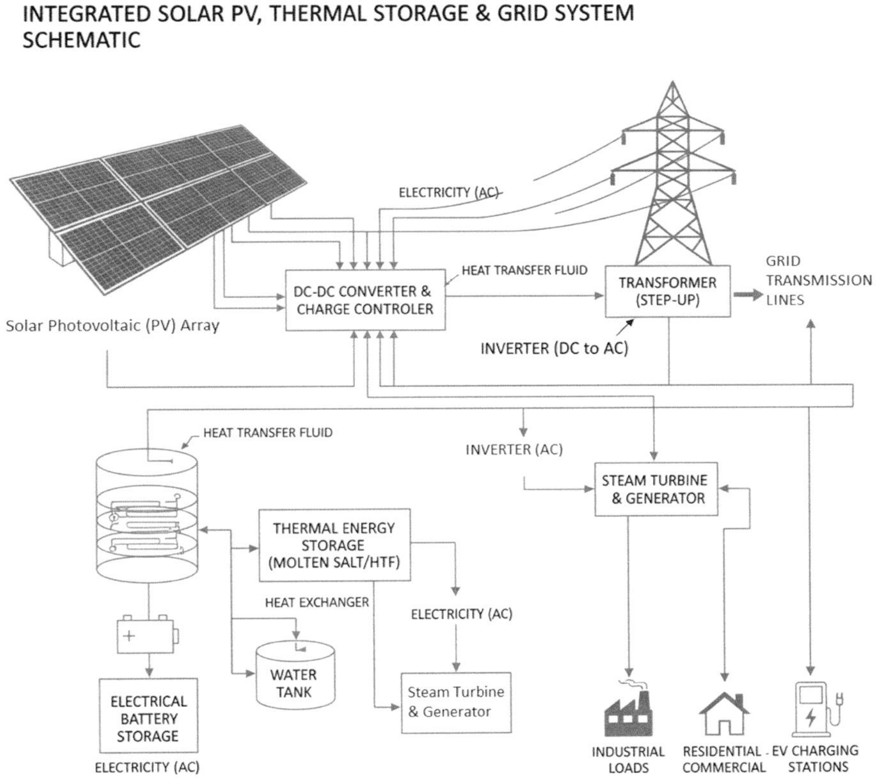

FIGURE 2.6
Integration of TES with solar PV system. (*Source*: Author.)

2.4.3 Integration of Energy Storage System with Biomass Energy

Thermal energy storage systems are integrated with biomass combined heat and power plants (CHPs). These plants simultaneously provide electricity and useful thermal energy (heat). The heat provided during the biomass conversion processes is then stored to provide a sustainable solution to the intermittency of this renewable energy technology. The heat is stored in TES to improve the load management and ensure a steady supply of heat even in the event of feedstock shortages. The thermal reliability is improved with the integration of sensible heat storage (such as pressurized water tanks) or latent storage using salt hydrates.

By helping biomass CHP manage load variations and fuel variability, TES improves annual efficiency by 10–15% [13]. A consistent district heating supply is ensured by integrating PCM-based TES. Figure 2.8 represents the integration of an energy storage system with a biomass CHP system.

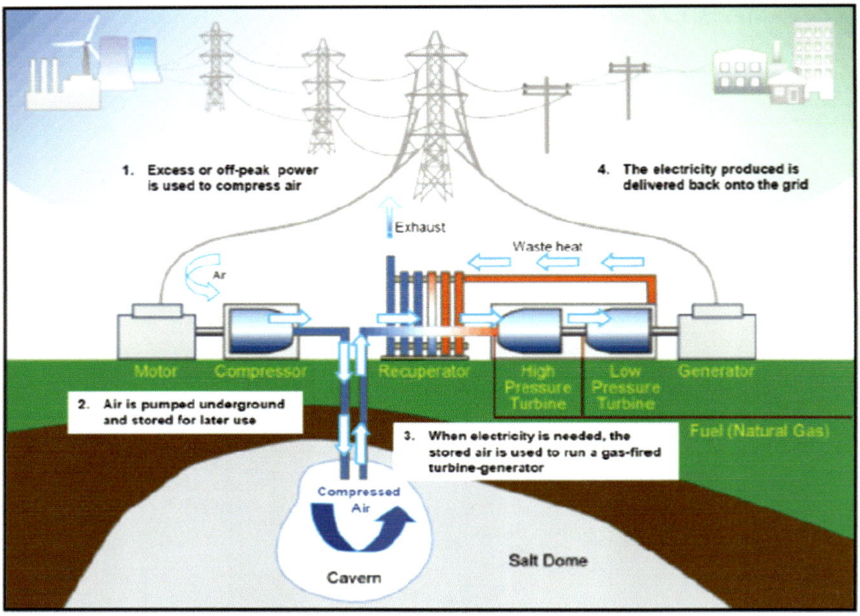

FIGURE 2.7
Schematic diagram of wind power energy system integrated with CAES. (From [12].)

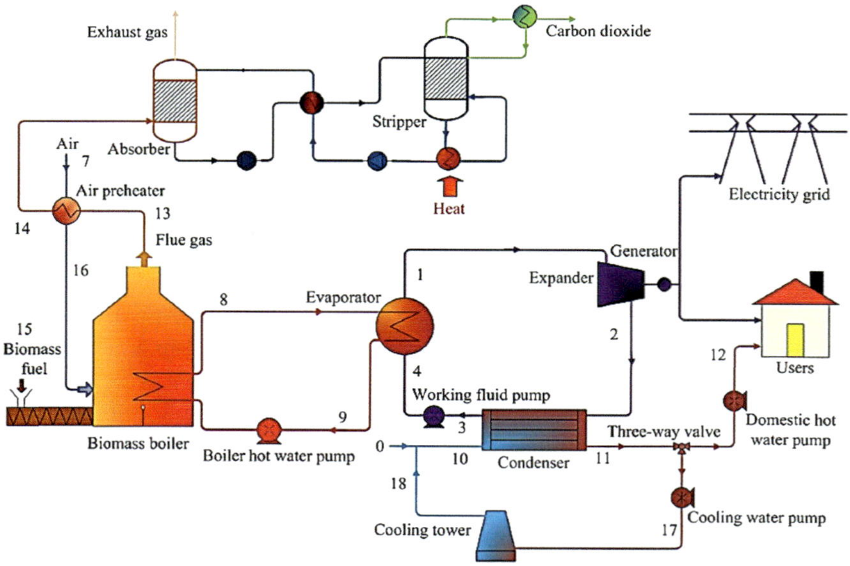

FIGURE 2.8
Schematic diagram of biomass energy system with integration of CHP system. (From [14].)

2.4.4 Integration of Energy Storage Systems with Geothermal Energy

Geothermal energy system integrated with thermal batteries to provide sustainable energy output. The underground reservoirs or aquifers are used as thermal batteries. The integration not only improves the overall efficiency of the system but also reduces the carbon emissions. Underground thermal energy storage (UTES), Aquifer thermal energy storage (ATES), and Enhanced geothermal systems (EGS) are examples of these integrated systems. All these systems use different technologies to convert an excess amount of energy into heat and back into a usable form of energy during peak hours. These systems are frequently combined with pumped thermal energy storage (PTES) systems to provide large-scale energy management by converting excess electricity into heat and back into electricity during peak hours [15]. A simplified thermal integration is represented here in Figure 2.9.

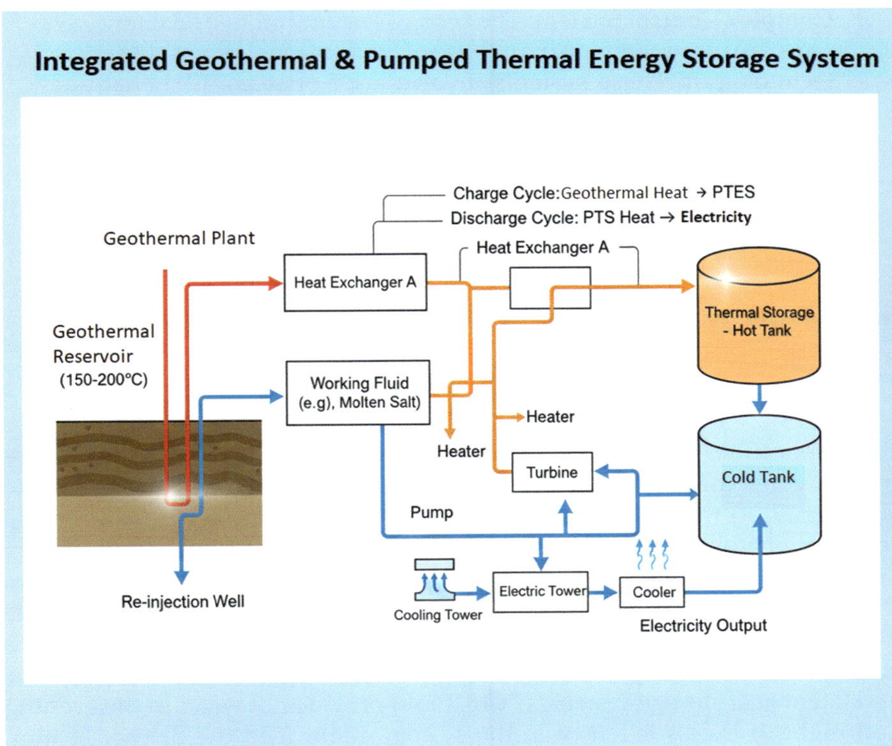

FIGURE 2.9
Geothermal energy integration with TES. (*Source*: Author.)

2.5 Challenges in Integration of Energy Storage Systems with RETs

The following are some of the challenges being faced during the integration of energy storage systems with renewable energy technologies.

- **Matching and Sizing:** The accuracy of sizing the thermal energy storage system is a challenging task; under-sizing lessens benefits, while oversizing wastes money.
- **Thermal Losses:** Thermal losses are associated with such a system, as inappropriate insulation often leads to the loss of thermal energy available or being stored.
- **Materials Cost:** Thermal energy storage materials are expensive and aren't available commercially. This makes the use of such materials limited and costly.
- **Complex Control System:** The control systems for hybrid energy systems are more complex than those of conventional renewable energy technologies. These systems involve a lot of steps, like charging/discharging control, management of multiple storage layers, and variability of renewable supply, etc.
- **Integration with current infrastructure:** Integration with existing power plants, industrial facilities, or HVAC systems is a challenging task as it requires more technical expertise.

2.6 Case Study: An Integrated System Consisting of Solar PV, Heat Pump, SHS, and LHS

The application of RET with TES integration at the building level is illustrated by this conceptual case.

Referring to the above mentioned Figure 2.10, during the day, a rooftop series of photovoltaic modules produces electricity. As there is a sufficient amount of electricity being produced during the day time therefore, the extra PV electricity powers a resistive heater or heat pump that charges a sensible heat storage system and a latent heat storage (PCM tank). The heat stored in a sensible heat storage system is used for space heating at a lower temperature, and latent heat storage systems (PCM Tank) provide hot water in the evening and night. If excess is left over, the system might feed the district heating network. This integrated system provides the following advantages,

- Management of building loads to lower evening grid demand.
- Utilizing the maximum power production of the photovoltaic system.

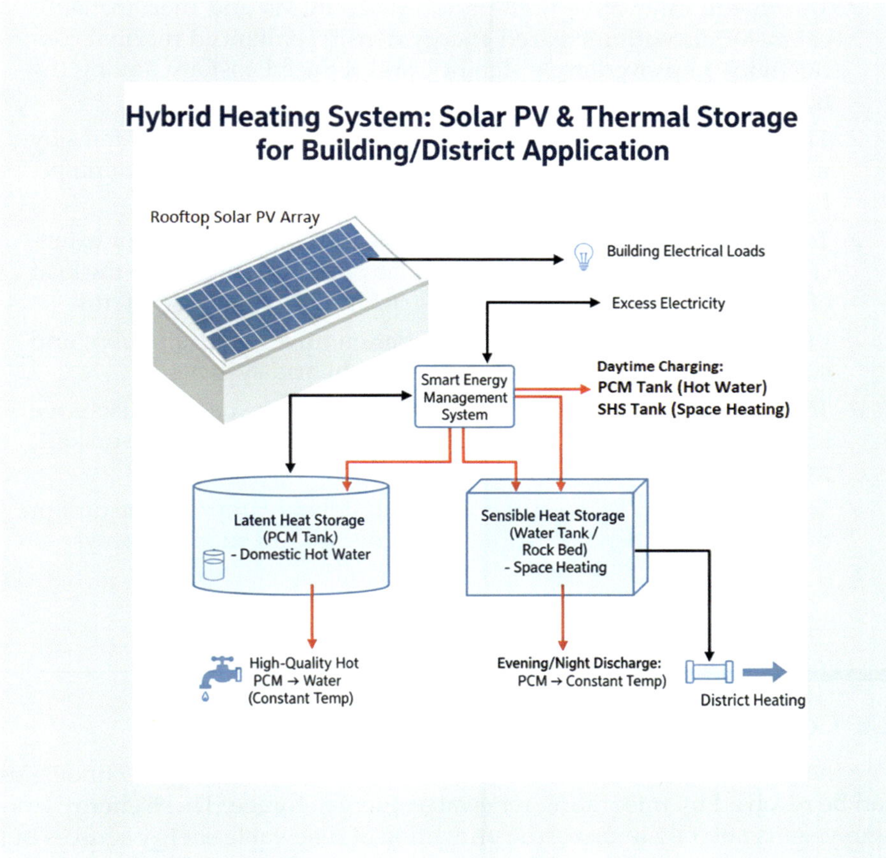

FIGURE 2.10
Illustration of case study. (*Source*: Author.)

- Provide a solution to the dependency on fossil fuels and grid electricity.
- The overall grid load is reduced.
- A combination of sensible and latent heat storage systems provides a better use of heat storage

2.7 Future Trends and Research Directions

The following are the future trends and research directions for the integration of energy storage systems with renewable energy technologies.

- ✓ To provide innovative high-performance PCMs and thermochemical materials with increased energy density, enhanced thermal conductivity, improved cycle stability, and reduced cost are among the field's primary future directions.
- ✓ The methods to improve the performance of such hybrid systems by using fins, encapsulation, composite media, and additive manufacturing techniques, etc.
- ✓ To design more sustainable seasonal heat storage systems for extensive renewable energy sources will also provide a solution to the load efficient load management, e.g., district heating/cooling systems.
- ✓ Providing techno-economic optimization, lifecycle evaluation, and sustainability analyses of these RET-TES hybrid systems.
- ✓ To make the system smart in control by real-time charge/discharge scheduling, TES operation optimization, and the use of sensors, IoT, predictive analytics, and artificial intelligence.
- ✓ To use the modern simulation tools will not only improve the design but also comprehend contributions to decarbonization pathways.

2.8 Conclusions

The intermittency problems associated with renewable energy technologies can be resolved by integrating renewable energy storage. These energy storage systems not only improve the utilization of renewable energy sources but also provide an alternative solution to the energy demands during off-peak hours. The hybridization of RETs with energy storage systems will provide more reliable, adaptable, and sustainable energy systems.

References

1. W. Shockley, and H. Queisser, "Detailed balance limit of efficiency of p–n junction solar cells," *Taylorfrancis.comW Shockley, H QueisserRenewable Energy, 2018•taylorfrancis.com*, vol. 2–4, pp. 35–54, 2018, doi: 10.4324/9781315793245-44/DETAILED-BALANCE-LIMIT-EFFICIENCY-P.
2. A. H. Ali, R. A. El-Kammar, H. F. A. Hamed, A. A. Elbaset, and A. Hossam, "Monitoring and enhancing the performance of PV systems using IoT and artificial intelligence algorithms," *ERU Research Journal*, vol. 3, no. 1, pp. 950–964, 2024, doi: 10.21608/ERURJ.2024.241998.1080.

3. Y. Khan, D. Singh, and S. Kumar, ... S. M.-J. of R., and undefined 2024, "Performance evaluation of the combined helium Brayton cycle and organic Rankine cycles for solar power tower application—A comparative study," *pubs. aip.org*.
4. M. Recalde, T. Woudstra, and P. V. Aravind, "Gasifier, solid oxide fuel cell integrated systems for energy production from wet biomass," *Frontiers in Energy Research*, vol. 7, p. 129, 2019, doi: 10.3389/FENRG.2019.00129/FULL.
5. C. Jiménez, et al., "Global intercomparison of 12 land surface heat flux estimates," *Wiley Online Library*, vol. 116, no. 2, 2011, doi: 10.1029/2010JD014545.
6. I. Sarbu, and C. Sebarchievici, "A comprehensive review of thermal energy storage," *Sustainability*, vol. 10, no. 1, p. 191, 2018, doi: 10.3390/SU10010191.
7. M. Jurczyk, et al., "Review of thermal energy storage materials for application in large-scale integrated energy systems—Methodology for matching heat storage solutions for given applications," *Energies*, vol. 17, no. 14, p. 3544, 2024, doi: 10.3390/EN17143544.
8. A. Sathishkumar, P. Sundaram, M. Cheralathan, and P. G. Kumar, "Effect of nano-enhanced phase change materials on performance of cool thermal energy storage system: A review," *Journal of Energy Storage*, vol. 78, p. 110079, 2024, doi: 10.1016/J.EST.2023.110079.
9. A. J. Carrillo, J. González-Aguilar, M. Romero, and J. M. Coronado, "Solar energy on demand: A review on high temperature thermochemical heat storage systems and materials," *Chemical Reviews*, vol. 119, no. 7, pp. 4777–4816, 2019, doi: 10.1021/ACS.CHEMREV.8B00315.
10. S. S. Mostafavi Tehrani, Y. Shoraka, K. Nithyanandam, and R. A. Taylor, "Cyclic performance of cascaded and multi-layered solid-PCM shell-and-tube thermal energy storage systems: A case study of the 19.9 MWe gemasolar CSP plant," *Applied Energy*, vol. 228, pp. 240–253, 2018, doi: 10.1016/J.APENERGY.2018.06.084.
11. E. Bazdar, M. Sameti, F. Nasiri, and F. Haghighat, "Compressed air energy storage in integrated energy systems: A review," *Renewable and Sustainable Energy Reviews*, vol. 167, p. 112701, 2022, doi: 10.1016/J.RSER.2022.112701.
12. L. Zhang, M. Xie, K. Ye, S. Li, and L. Chen, "Compressed air energy storage based on variable-volume air storage: A review," *Journal of Energy Storage*, vol. 110, p. 115361, 2025, doi: 10.1016/J.EST.2025.115361.
13. K. Lepiksaar, V. Mašatin, E. Latõšov, A. Siirde, and A. Volkova, "Improving CHP flexibility by integrating thermal energy storage and power-to-heat technologies into the energy system," *Smart Energy*, vol. 2, p. 100022, 2021, doi: 10.1016/J.SEGY.2021.100022.
14. M. A. Bagherian, K. Mehranzamir, S. Rezania, Z. Abdul-Malek, A. B. Pour, and S. M. Alizadeh, "Analyzing utilization of biomass in combined heat and power and combined cooling, heating, and power systems," *Processes*, vol. 9, no. 6, p. 1002, 2021, doi: 10.3390/PR9061002.
15. A. Vannoni, and A. Sorce, "Electrical storage or gas power plant flexibility by thermal storage? A market analysis," *Journal of Energy Storage*, vol. 126, p. 116823, 2025, doi: 10.1016/J.EST.2025.116823.

3

Low, Medium, and High Temperatures Thermal Energy Storage (TES) Methods

Laveet Kumar and Bilawal A. Bhayo

3.1 Introduction

The transition toward a low-carbon energy future has placed energy storage technologies at the center of global research, development, and deployment efforts. Renewable energy sources such as solar and wind are inherently intermittent and variable, producing energy that does not always coincide with demand patterns. Without adequate storage solutions, this mismatch can result in wasted energy, grid instability, and continued reliance on fossil fuel-based backup systems [1]. While electrochemical batteries and mechanical storage solutions are widely discussed, thermal energy storage (TES) offers unique advantages in situations where heat, rather than electricity, is the primary energy demand. TES can reduce reliance on fossil fuels in heating and industrial sectors, improve the efficiency of combined heat and power systems, and enable dispatchable renewable electricity generation coupled with high-temperature processes [2]. At its core, TES involves capturing thermal energy at one point in time and making it available for use at another. The energy may be stored as sensible heat, latent heat, or in the form of chemical potential energy. Sensible heat storage relies on raising the temperature of a medium such as water, oil, or solid materials. Latent heat storage harnesses the energy absorbed or released during phase transitions, such as melting and solidification. Thermochemical storage uses reversible chemical reactions, offering exceptionally high energy density and long-term storage potential with minimal heat losses [3].

TES systems are broadly classified by their operating temperature ranges. Low-temperature storage (<100°C) is particularly useful in residential heating, cooling, and refrigeration applications. Medium-temperature storage (100–400°C) plays an important role in industrial processes, cogeneration, and solar thermal plants. High-temperature storage (>400°C) is essential in concentrated solar power (CSP), heavy industry, and large-scale electricity generation. Each category has distinct materials, technologies, advantages, and challenges [4]. Figure 3.1 illustrates the classification of thermal

DOI: 10.1201/9781003516132-3

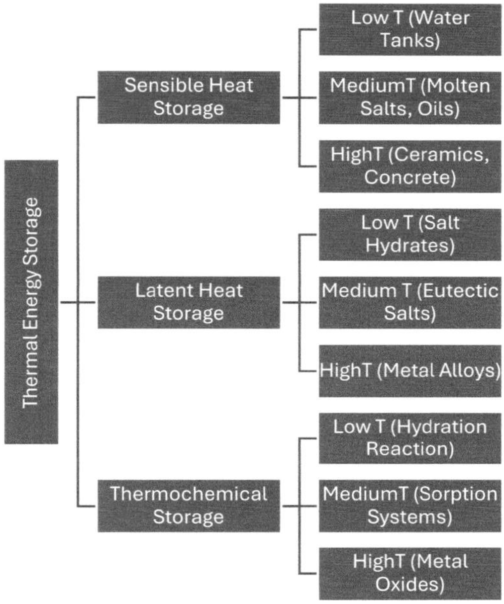

FIGURE 3.1
Classification of thermal energy storage by type and temperature ranges.

energy storage (TES) technologies according to both storage mechanism and operating temperature range. At the highest level, TES can be divided into three categories: sensible heat storage, latent heat storage, and thermochemical storage. Each of these mechanisms can operate across low, medium, and high temperature ranges, with specific storage media and applications associated with each. For sensible heat storage, water tanks dominate in low-temperature applications, while molten salts and thermal oils are widely used at medium temperatures, and solid media such as ceramics and concrete are preferred for high-temperature systems. Latent heat storage relies on phase change materials (PCMs), with paraffins and salt hydrates suitable for low-temperature use, eutectic salts for medium ranges, and specialized alloys for high temperatures. Thermochemical storage includes hydration reactions at low temperatures, sorption-based systems at medium levels, and reversible metal oxide reactions for high-temperature, high-density storage. This classification provides a clear framework for understanding how TES technologies are matched to different applications depending on their operating conditions, energy density, and material suitability [4].

Sensible heat storage involves raising or lowering the temperature of a storage medium, such as water, oils, rocks, or ceramics. The stored energy is directly proportional to the mass of the medium, its specific heat capacity, and the temperature difference [5]. Latent heat storage is based on the energy absorbed or released during the phase change of a material, usually from solid to liquid or vice versa. Phase change materials (PCMs) such as paraffins, fatty acids, and salt hydrates are typically employed. The main advantage is the ability to store large amounts of heat at nearly constant temperature, governed by the latent heat of fusion [6, 7]. Thermochemical storage relies on reversible chemical reactions, where energy is stored in the form of chemical bonds. Examples include hydration–dehydration cycles, adsorption processes, and metal oxide redox reactions. The amount of energy stored is determined by the enthalpy change of the reaction. Thermochemical systems offer exceptionally high storage densities and the possibility of long-term or seasonal storage with minimal losses, though they are still in the developmental stage compared to sensible and latent storage [8]. Together, these three principles form the foundation of TES technologies, and their selection depends on the temperature range, energy density requirements, and application context.

This chapter provides a detailed examination of TES technologies across these three temperature ranges, offering insights into their design, applications, comparative strengths, and future potential.

3.2 Low-Temperature Thermal Energy Storage (LTES)

3.2.1 Storage Mechanisms

Low-temperature thermal energy storage is primarily based on sensible heat and latent heat mechanisms. Sensible heat storage systems are the simplest and most mature, relying on increasing the temperature of a medium such as water, soil, or packed beds of pebbles and gravel. Water is particularly effective due to its high specific heat capacity, affordability, and widespread availability. Large, insulated water tanks are widely employed in residential and district heating systems to provide hot water on demand. Latent heat storage systems rely on phase change materials (PCMs) that absorb or release large amounts of heat at nearly constant temperatures during melting and solidification. PCMs such as paraffin waxes, fatty acids, and salt hydrates are commonly used in building applications to regulate indoor temperature and improve thermal comfort. Their ability to maintain stable operating temperatures makes them particularly valuable in refrigeration and cold storage systems [9].

Another widespread approach is the use of chilled water and ice storage, especially in commercial HVAC systems. In these systems, chillers produce ice or chilled water during off-peak hours when electricity is cheaper or demand is low. The stored cold is then used during peak hours, reducing electricity costs and easing pressure on the grid [10].

3.2.2 Applications and Limitations

Low-temperature TES is extensively applied in buildings and urban infrastructure. In residential and commercial buildings, hot-water tanks and PCM-based systems provide heating and cooling support, while reducing reliance on electricity at peak times. District heating and cooling networks also benefit from large-scale hot water tanks, which store excess thermal energy generated by combined heat and power plants or renewable sources [11]. Refrigeration and cold storage are another important application area. By integrating PCMs and ice storage, food supply chains, supermarkets, and pharmaceutical storage facilities can ensure stable temperatures while significantly reducing electricity demand during peak hours [10, 12].

Low-temperature thermal energy storage (LTES) is most commonly applied in the building sector, where it provides heating, cooling, and domestic hot water services. The simplest and most widespread approach is the use of insulated hot-water tanks, which store thermal energy from boilers, heat pumps, or solar collectors. These systems allow heat to be stored during off-peak times or periods of renewable generation and used when demand peaks [10]. Another method is the incorporation of phase change materials (PCMs) into building envelopes, such as wallboards or ceiling panels. These materials absorb heat when room temperatures rise above comfort levels and release it when temperatures fall, thereby stabilizing indoor climates and reducing heating or cooling loads. For cooling applications, ice storage systems are widely implemented in commercial buildings. Ice is produced at night when electricity is cheaper and stored in large insulated tanks. During the day, the stored ice is melted to provide cooling, thereby reducing electricity consumption and peak demand charges [10, 12]. Together, these approaches illustrate the versatility of LTES in the built environment. By integrating sensible and latent heat storage systems, buildings can significantly improve their energy efficiency, lower operational costs, and contribute to reducing peak electricity demand.

The main advantages of LTES include its low cost, high technological maturity, and ease of implementation. Systems based on water storage, for example, are widely deployed and well understood. They require relatively simple infrastructure and minimal operational complexity. However, LTES systems have limited energy density, which means large volumes are needed to store significant amounts of energy. Moreover, their applicability is restricted to low-grade heating and cooling, limiting their usefulness in industrial or high-temperature energy-intensive applications.

3.3 Medium-Temperature Thermal Energy Storage (MTES)

3.3.1 Storage Mechanisms

Medium-temperature TES systems are typically based on sensible heat storage using pressurized water, thermal oils, and molten salts. Pressurized water systems operate up to 200–250°C but require reinforced vessels and safety mechanisms. Synthetic oils can operate at temperatures up to 400°C and are often used in solar thermal plants. Molten nitrate salts, commonly a mixture of sodium nitrate and potassium nitrate, are another widely used medium, particularly in solar thermal applications due to their stability and relatively low cost. Latent heat storage in the medium range makes use of eutectic salts and specialized PCMs, which can absorb and release large amounts of heat in narrow temperature bands. These are particularly useful in industrial processes that require stable process temperatures. Thermochemical storage systems also emerge in this range, employing chemical sorption or reversible reactions that allow energy to be stored without significant thermal losses over long durations [13, 14].

3.3.2 Applications and Limitations

Medium-temperature TES has broad applications across industry and energy systems. Industrial sectors such as textiles, food processing, chemicals, and paper manufacturing require process heat between 150°C and 300°C, which can be efficiently supplied using MTES systems. TES also facilitates the recovery of industrial waste heat, improving energy efficiency and reducing emissions. In combined heat and power (CHP) systems, TES allows for greater operational flexibility by decoupling electricity generation from heat supply. By storing excess thermal energy, CHP systems can adjust their output to match demand more effectively [15].

Solar thermal power plants, particularly those using parabolic trough collectors, also rely on medium-temperature TES. By storing solar heat collected during the day, these plants can extend their operating hours into the night, thereby improving reliability and reducing dependence on fossil fuels. Medium-temperature thermal energy storage (MTES) systems are typically integrated into parabolic trough solar thermal plants, which operate within the 150–400°C range. The integration of TES allows the plant to decouple solar energy collection from electricity generation, enabling electricity supply during evening peaks and improving grid reliability [16].

Medium-temperature TES faces several challenges. The thermal stability of oils and molten salts is a concern, as degradation over time can limit system lifetimes. Heat losses also become more significant at these temperature ranges, requiring high-performance insulation. Furthermore, retrofitting industrial plants with TES can be economically challenging, as the initial investment costs are high and integration into existing processes may be complex [17].

3.4 High-Temperature Thermal Energy Storage (HTES)

3.4.1 Storage Mechanisms

At high temperatures above 400°C, TES systems require robust and durable storage media. Molten salts such as chlorides and carbonates are widely used in concentrated solar power (CSP) plants due to their ability to store large amounts of energy and operate at high temperatures. Solid media, such as concrete, ceramics, and graphite, are increasingly being studied for their potential in high-temperature TES. These materials offer high thermal stability, mechanical strength, and relatively low cost, making them promising for large-scale applications [18]. Thermochemical storage systems are particularly attractive at high temperatures. They employ reversible chemical reactions—such as those involving metal oxides that can store energy in chemical bonds. These systems have the potential for seasonal storage and can achieve very high storage densities with minimal energy losses during idle periods [19, 20].

3.4.2 Applications and Limitations

High-temperature TES plays a central role in concentrated solar power plants, where solar energy is concentrated onto receivers and stored in molten salts. This stored energy is then used to generate electricity long after sunset, providing dispatchable renewable power and enabling baseload operation. In heavy industries such as metallurgy, glass, and cement manufacturing, TES systems are being explored to provide process heat at temperatures above 1000°C. This offers a pathway to decarbonize some of the most energy-intensive and carbon-heavy sectors of the global economy. High-temperature TES also has potential in grid-scale electricity storage, particularly when integrated with power cycles such as Rankine or Brayton cycles. By storing energy as heat and later converting it back to electricity, HTES can complement batteries by providing long-duration storage capabilities [21].

High-temperature thermal energy storage (HTES) is most prominently deployed in CSP tower plants, where operating temperatures typically exceed 500°C and can reach up to 1000°C. These systems include Heliostat Field – Thousands of computer-controlled mirrors (heliostats) track the sun and focus solar radiation onto a central receiver located at the top of a tower. Central Receiver – A receiver absorbs concentrated sunlight and transfers the thermal energy to a working fluid, typically molten salt (commonly a binary mixture of sodium nitrate and potassium nitrate). This fluid is heated to high temperatures, often around 565°C. Hot and Cold Storage Tanks – Like parabolic trough systems, tower plants use a two-tank direct storage configuration. The hot tank stores molten salt heated in the receiver, while the cold tank holds the cooler return salt. This allows for continuous circulation and stable operation [22, 23].

HTES systems face significant technical and economic barriers. Corrosion and material degradation are major concerns at elevated temperatures, especially for molten salt systems. Solid materials may experience thermal stresses and cracking, which can reduce system reliability. Advanced insulation materials are necessary to minimize heat losses, but add to the cost. Consequently, HTES systems are currently more expensive than low- and medium-temperature options, although their potential impact on renewable integration and industrial decarbonization is much greater.

3.5 Comparative Analysis

When comparing TES technologies, it is clear that each temperature range serves distinct purposes. Low-temperature systems are well-suited for residential and commercial applications, medium-temperature systems address industrial needs, and high-temperature systems enable large-scale renewable integration and heavy industry decarbonization. The selection of a thermal energy storage (TES) technology is highly dependent on the temperature range, energy density, application requirements, and economic considerations. Table 3.1 summarizes the main characteristics of low-, medium-, and high-temperature TES systems. Overall, this comparative perspective shows that no single TES system is universally superior. Instead, the choice of technology is application-driven, depending on the thermal requirements, energy density, system scale, and cost considerations. LTES is most effective for buildings and cold storage, MTES is increasingly relevant for industrial heat and solar thermal, and HTES is essential for large-scale renewable integration and industrial decarbonization.

3.6 Conclusions and Future Trends

Thermal energy storage is an indispensable enabler of a sustainable energy future. Low-temperature systems provide affordable and reliable solutions for residential heating and cooling, medium-temperature systems enhance industrial efficiency and solar thermal power generation, while high-temperature systems offer the promise of decarbonizing heavy industry and delivering dispatchable renewable electricity.

Although challenges remain in terms of material stability, cost, and system integration, ongoing research and supportive policy frameworks are likely to overcome these barriers. By tailoring TES technologies to specific temperature ranges and applications, TES can significantly contribute to global decarbonization, grid stability, and energy security.

TABLE 3.1

Comparative Characteristics of Low-, Medium-, and High-Temperature TES Systems

Parameter	LTES (<100°C)	MTES (100–400°C)	HTES (>400°C)
Materials	Water, chilled water, ice, paraffins, salt hydrates, fatty acids	Molten salts (nitrates), synthetic oils, eutectic salts, medium-T PCMs	Molten salts (chlorides, carbonates), ceramics, concrete, graphite, metal oxides (thermochemical)
Storage Density	Low: ~20–50 kWh/m^3	Medium: ~50–150 kWh/m^3	High: >150 kWh/m^3; thermochemical >250 kWh/m^3
Applications	Domestic hot water, HVAC, refrigeration/cold storage, district heating & cooling	Industrial process heat (food, textiles, chemicals, paper), CHP, solar thermal plants	Concentrated Solar Power (CSP), metallurgy, cement & glass industries, long-duration electricity storage
Maturity	High (commercial, widespread)	Medium (commercial, but limited scale and integration challenges)	Emerging/Developing (demonstration to pilot stage; CSP is the most advanced application)
Cost	Low (USD 10–30/kWh_t)	Moderate (USD 30–60/kWh_t)	High (USD 60–100+/kWh_t, depending on technology)
Advantages	Low cost, mature, simple operation, widely available materials	Higher operating temperature, suitable for industry and solar thermal, flexible scale	High storage density, long-duration potential, enables dispatchable renewables, supports heavy industry
Limitations	Low energy density, limited to low-grade heat	Stability of oils/salts, thermal losses, high insulation needs, and economic challenges	Corrosion, thermal stresses, high capital cost, complex system integration

Source: Refs [3, 4, 24, 25].

The future of TES will be shaped by advances in materials, hybridization, and system integration. Researchers are developing nano-enhanced PCMs with improved thermal conductivity and encapsulated composites that prevent material leakage during phase transitions. High-performance thermochemical storage systems are being designed for seasonal storage, potentially revolutionizing heating and cooling in colder regions. Hybrid TES systems that combine sensible, latent, and thermochemical storage mechanisms offer the possibility of achieving the best attributes of each approach, balancing cost, energy density, and operating flexibility. Integration of TES with renewable energy systems, district heating networks, and green hydrogen production pathways will further increase their role in sustainable energy systems. From a policy perspective, economic incentives, carbon pricing, and supportive regulations will play an essential role in accelerating TES adoption.

Without policy intervention, the high upfront costs of TES may hinder widespread deployment, despite its long-term benefits.

References

1. M. Urf Manoo, F. Shaikh, L. Kumar, and M. Arıcı, "Comparative techno-economic analysis of various stand-alone and grid connected (solar/wind/fuel cell) renewable energy systems," *International Journal of Hydrogen Energy*, vol. 52, pp. 397–414, 2024. https://doi.org/10.1016/j.ijhydene.2023.05.258
2. P. Arévalo, D. Ochoa-Correa, and E. Villa-Ávila, "Advances in thermal energy storage systems for renewable energy: A review of recent developments," *Processes*, vol. 12, no. 9, pp. 1–29, 2024, doi: 10.3390/pr12091844.
3. M. A. Rosen, *Thermal Energy Thermal Energy*. 2011.
4. A. K. Dewangan, S. Q. Moinuddin, M. Cheepu, S. K. Sajjan, and A. Kumar, "Thermal energy storage: Opportunities, challenges and future scope," in *Thermal Energy Systems*, Taylor & Francis, 2023, pp. 17–28.
5. C. Amy, C. C. Kelsall, A. LaPotin, M. Pishahang, and A. Henry, "Chapter 3 – Ultrahigh temperature sensible heat storage and heat transfer fluids," in *Woodhead Publishing Series in Energy, Ultrahigh Temperature Thermal Energy Storage, Transfer and Conversion*, A. Henry, Ed., Woodhead Publishing, 2021, pp. 57–84. https://doi.org/10.1016/B978-0-12-819955-8.00003-X
6. M. D. S. Hossain, L. Kumar, A. Arshad, J. Selvaraj, A. K. Pandey, and N. A. Rahim, "A comparative investigation on solar PVT- and PVT-PCM-based collector constancy performance," *Energies*, vol. 16, no. 5, 2023, doi: 10.3390/en16052224.
7. H. T. Awan, L. Kumar, W. P. Wong, R. Walvekar, and M. Khalid, "Recent progress and challenges in MXene-based phase change material for solar and thermal energy applications," *Energies*, vol. 16, no. 4, 2023, doi: 10.3390/en16041977.
8. R. Salgado-Pizarro, A. Calderón, A. Svobodova-Sedlackova, A. I. Fernández, and C. Barreneche, "The relevance of thermochemical energy storage in the last two decades: The analysis of research evolution," *Journal of Energy Storage*, vol. 51, p. 104377, 2022. https://doi.org/10.1016/j.est.2022.104377
9. M. A. Khuda, L. Yan, N. Sarunac, and C. Romero, "Design, analysis, and testing of a prototype-scale latent heat thermal energy storage (LTES) system," *Journal of Energy Storage*, vol. 89, p. 111796, 2024. https://doi.org/10.1016/j.est.2024.111796
10. H. Ayaz, V. Chinnasamy, J. Yong, and H. Cho, "Review of technologies and recent advances in low-temperature sorption thermal storage systems," *Energies*, vol. 14, no. 19, 2021, doi: 10.3390/en14196052.
11. Y. R. Fang, et al., "Performance and economic analysis of photovoltaic/thermal systems with phase change materials and a parallel serpentine design in dusty conditions," *Applied Thermal Engineering*, vol. 268, p. 125890, 2025. https://doi.org/10.1016/j.applthermaleng.2025.125890
12. M. Shahzaib, A. Moeez, A. G. Memon, and L. Kumar, "Parametric based techno-economic evaluation for a solar thermal-PV integrated multi-commodity storage facility," *Energy Storage*, vol. 6, no. 6, p. e70022, 2024. https://doi.org/10.1002/est2.70022

13. W.-D. Steinmann, *Thermal energy storage for medium and high temperatures.* Springer, 2022.
14. T. R. Gbenou, A. Fopah-Lele, and K. Wang, "Recent status and prospects on thermochemical heat storage processes and applications," *Entropy*, vol. 23, no. 8, 2021, doi: 10.3390/e23080953.
15. C. Chen, Z. Ge, and Y. Zhang, "Study of combined heat and power plant integration with thermal energy storage for operational flexibility," *Applied Thermal Engineering*, vol. 219, p. 119537, 2023. https://doi.org/10.1016/j.applthermaleng.2022.119537
16. D. S. Jayathunga, H. P. Karunathilake, M. Narayana, and S. Witharana, "Phase change material (PCM) candidates for latent heat thermal energy storage (LHTES) in concentrated solar power (CSP) based thermal applications – A review," *Renewable and Sustainable Energy Reviews*, vol. 189, p. 113904, 2024. https://doi.org/10.1016/j.rser.2023.113904
17. K. Kant, and R. Pitchumani, "Advances and opportunities in thermochemical heat storage systems for buildings applications," *Applied Energy*, vol. 321, p. 119299, 2022. https://doi.org/10.1016/j.apenergy.2022.119299
18. M. Opolot, C. Zhao, M. Liu, S. Mancin, F. Bruno, and K. Hooman, "A review of high temperature (≥ 500°C) latent heat thermal energy storage," *Renewable and Sustainable Energy Reviews*, vol. 160, p. 112293, 2022. https://doi.org/10.1016/j.rser.2022.112293
19. L. Miró, E. Oró, D. Boer, and L. F. Cabeza, "Embodied energy in thermal energy storage (TES) systems for high temperature applications," *Applied Energy*, vol. 137, pp. 793–799, 2015. https://doi.org/10.1016/j.apenergy.2014.06.062
20. Q. Wang, et al., "A review of eutectic salts as phase change energy storage materials in the context of concentrated solar power," *International Journal of Heat and Mass Transfer*, vol. 205, p. 123904, 2023. https://doi.org/10.1016/j.ijheatmasstransfer.2023.123904
21. Z. Li, et al., "Applications and technological challenges for heat recovery, storage and utilisation with latent thermal energy storage," *Applied Energy*, vol. 283, p. 116277, 2021. https://doi.org/10.1016/j.apenergy.2020.116277
22. B. Hernández, N. L. Ferber, M. Abdullah, A. Mayyas, N. Calvet, and M. Chiesa, "Central receiver-based concentrated solar power plants part 1: A historical review," *Solar Energy*, vol. 299, p. 113693, 2025. https://doi.org/10.1016/j.solener.2025.113693
23. B. Hernández, N. L. Ferber, M. Abdullah, A. Mayyas, N. Calvet, and M. Chiesa, "Central receiver-based CSP plants part 2: Components categorization and future prospects," *Solar Energy*, vol. 299, p. 113740, 2025. https://doi.org/10.1016/j.solener.2025.113740
24. L. F. Cabeza, "2 - Advances in thermal energy storage systems: Methods and applications," in *Woodhead Publishing Series in Energy*, (Second Edition), L. F. Cabeza, Ed., Woodhead Publishing, 2021, pp. 37–54. https://doi.org/10.1016/B978-0-12-819885-8.00002-4
25. A. Khademi, et al., "A brief review on different hybrid methods of enhancement within latent heat storage systems," *Journal of Energy Storage*, vol. 54, p. 105362, 2022. https://doi.org/10.1016/j.est.2022.105362

4

Research Status and Development Trends on Cold Energy Storage Technology and Its Engineering Applications

Mohammad Hamid, Song Mengjie, Aye Aye Thet, Keke Shao, Yonghui Liang, and Sirui Yu

4.1 Introduction

Cold energy storage (CES) has been identified as a key technology to address the growing need for cooling in various engineering disciplines, including buildings, industrial applications, cold storage, hospitals, and data centers. The need for cooling is related to peak electricity demand. As cooling systems operate mainly during peak hours or days. Therefore, this results in increased energy costs, fossil fuel-based peak power generation, and environmental impact. Cold energy storage systems produce cooling energy during off-peak hours, characterized by lower electricity demand and prices, and then store it to be used during peak hours. It is essential to note that cold energy storage promises enhanced system performance by reducing peak electrical demands, improving the grid, and reducing the necessary power generation and distribution system size. Further, cold energy storage promises actual economic benefits, such as lower operating costs and greater utilization of the existing equipment. There are various types of cold energy storage, including sensible cold storage, latent cold storage using phase change materials, and thermochemical cold storage. These types have varying storage densities, operating temperatures, and complexity. Cold energy storage is crucial in the development of sustainable energy systems. It allows the easier integration of various types of renewable energy sources, controls waste of cold energy with reduced greenhouse gas emissions during peak electricity generation. Therefore, cold energy storage is referred to as a potential key player in the provision and development of efficient, reliable, and sustainable cooling technologies and engineering systems.

The transition towards green energy through the optimization of renewable energy sources offers significant opportunities for reducing greenhouse gas emissions, which significantly impacts climate change and fosters sustainable development. Duration and capacity are the main limitations of

DOI: 10.1201/9781003516132-4

energy storage systems. Sustainable energy storage systems employ renewable energy sources either directly or by integrating energy from various sustainable resources. The primary objective of an energy storage system is to ensure cost-effectiveness, reliability, and accessibility for stakeholders. In addition, the equipment and processes utilized in the energy storage system must be sustainable. Various instruments are available to assess the sustainability of these systems, including life cycle assessment, key performance indicators, and process parameters. These tools are based on the fundamental principles of sustainability, encompassing environmental, economic, and social dimensions. Figure 4.1 illustrates the categorization of energy storage systems. Studies indicate that in 2022, approximately 32% of global food production was lost, with significant waste occurring at various stages of the supply chain. An in-depth analysis indicates that 13% of the total waste occurs at the earlier stages of the supply chain, specifically from post-harvest to the pre-retail phase. The remaining 19% is distributed across various end-use sectors, including commercial transactions, the food industry, and domestic waste.

The annual loss of food amounts to a significant 1.05 billion tons, reaching an average of 132 kg per person each year. Residential waste constitutes the largest segment of this total, accounting for 79 kg per person annually. Additionally, 36 kg is attributed to the food service sector, while 17 kg originates from the retail sector. Consequently, various methods are used to preserve food quality and extend its shelf life. These preservation techniques range from traditional methods such as sun-drying and smoking to modern

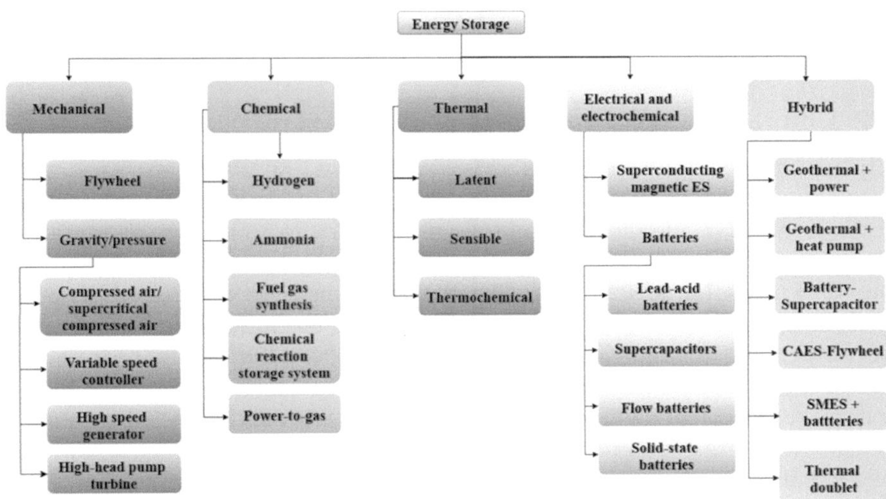

FIGURE 4.1
Classification of energy storage systems.

approaches, including coating, canning, refrigeration, freezing, solar dryer systems, and the use of food additives. Temperature is a critical factor in maintaining the quality of perishable products. As a result, refrigeration systems are utilized worldwide and play an essential role in the food supply chain. Various studies indicate it is estimated that there are around 0.05 million cold rooms, 5 million refrigeration vehicles, 1.2 million containers, 2000 million domestic refrigerators [1], and roughly 120 million commercial refrigerators. This study provides an in-depth examination of cold energy storage systems and the utilization of ice in these applications.

The present chapter offers a comprehensive overview of cold energy storage systems with regard to their classification, principle of operation, and engineering applications. A comprehensive literature survey was conducted using prominent scientific databases such as Web of Science, Scopus, and Google Scholar. Relevant research and review articles related to cold energy storage technologies from the available literature were selected and analyzed. This chapter is designed to address cold energy storage technologies in a systematic and comprehensive way. It begins with cold thermal energy storage technologies, such as sensible and latent cold storage technologies. It also covers phase change materials and dynamic ice storage technologies. It continues with cold energy storage technologies' main application domains, such as active and passive cold storage, electricity storage such as liquid air energy storage and pumped thermal energy storage, cold energy recovery from waste, and artificial intelligence-based cold energy storage technologies. Through a critical analysis of the available literature, this chapter aims to provide a comprehensive overview of cold energy storage technologies and their characteristics.

4.2 Cold Storage Technologies

Chilled warehouses utilize methods to maintain low temperatures through cold storage techniques and artificial cooling systems. These methods effectively address challenges related to inconsistency in cold energy in terms of their time and magnitude. Such systems are typically used to preserve various agricultural products, including vegetables, fruits, flowers, aquatic products, and meat. It is crucial to highlight that the operation and maintenance of cold storage during transit are essential for extending the shelf life of agricultural products. Cold energy storage technology has its roots in the fundamental research carried out on latent heat storage during the 1970s, a period when the energy crisis caused by the interruption of petroleum supplies necessitated the search for alternative energy management strategies. Initial research proved that phase change materials could store and deliver much more energy per unit volume than sensible heat storage systems using water or any other working fluids. Thus, leading to further research on salt hydrate and paraffin wax systems for both heating and cooling purposes. Practical aspects of cold storage technology were further enhanced by researchers in Japan during the 1980s and 1990s. Research on ice-based systems for cooling buildings led to the possibility of delaying the cooling load until nighttime hours when the price of electricity is relatively lower.

Significant contributions were made by European research communities in the 1990s and 2000s towards the development of salt hydrate technology in building applications and district cooling systems that have the capability of incorporating thermal storage. Investigations identified the challenges associated with the application of such technology, including thermal losses in long piping networks, integration challenges, and the need for maintenance of phase-change materials over long time periods. The simultaneous development of advanced control systems and energy management systems has provided opportunities for the implementation of more refined cold storage operations. There has been the development of optimization algorithms that adjust the charging or discharging patterns based on weather forecasts, electricity rates, and thermal demands.

4.2.1 Cold Thermal Energy Storage

Cold thermal energy storage technology involves the retention of cold energy within a medium that maintains a temperature lower than the typical ambient temperature or the device's operational temperature. Cold storage works on the principle of storing cold energy during off-peak periods and then supplying it during peak hours. In general, CES technology uses a vessel containing preservation material that enables cold transfer through the thermal exchanger. Integrating CES technology into existing cooling systems, such as refrigeration and air-conditioning systems, is regarded as highly cost-effective, energy-efficient, and reliable. Cooling systems are classified into two primary categories: passive and active systems. Active cooling systems are characterized by their substantial energy consumption, for example, refrigeration systems that produce cold air for cooling purposes. In contrast, passive cooling methods are distinguished by their reliance on temperature and pressure differentials to achieve cooling [2].

The advantages of CES in active cooling systems are primarily derived from two key factors: I) The integration of CES within the cooling system, specifically on the evaporator side, and the application terminal facilitates the storage of cold during periods of reduced prices of electricity or when a sufficient cold source is available. This system can subsequently deliver cooling based on user requirements, ensuring adaptability in cooling solutions for users II) The integration of CES with the cooling system, particularly on the condensation chamber side and a natural cold source, enhances the efficiency of the cold production system. This configuration provides an additional stable cold source when the available cold energy is insufficient [3]. The combination of CES with a refrigeration system is presented in Figure 4.2.

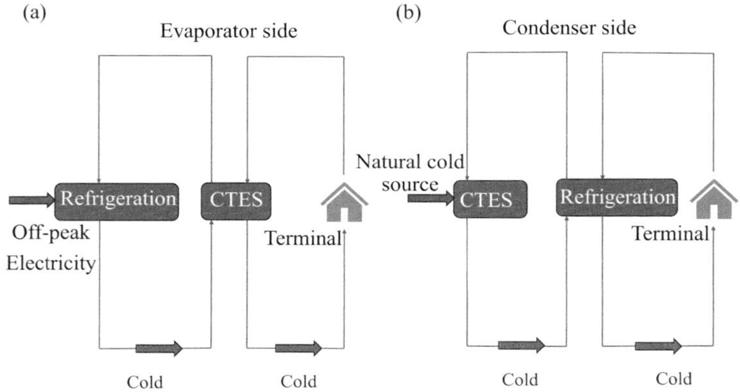

FIGURE 4.2
Combination of CES with refrigeration system.

Cold energy sources are typically derived from the surrounding ambient cold environment, electrical power, or the cold energy from LNG terminals. During periods of low electricity costs, cold energy is produced through refrigeration, air conditioning, and related systems, and subsequently stored in a cold storage unit for use during peak demand periods. This cold storage technique has the potential to substantially reduce electricity consumption during peak periods, thereby decreasing the typical cost of electricity. CES technology is widely applicable across various sectors, including industrial, manufacturing, and residential uses. It is utilized for space cooling, industrial process cooling, food preservation, and transportation within the cold chain.

Cold storage materials can be classified into two main categories: sensible heat materials and latent heat materials. Materials that utilize sensible heat capacity for the storage of cold energy are categorized as sensible heat materials. For instance, cold water is often selected as the medium, which demonstrates the advantages of low-temperature liquid water, high specific heat capacity, which is 4.18 kJ kg^{-1} K^{-1}, non-toxic properties, and cost-effectiveness. The use of rocks in natural environments offers a feasible option for large-scale cold storage solutions. Phase change materials can store or transfer cold energy through latent heat processes [4]. The substantial energy density associated with phase change enthalpy, along with the consistent temperature at which a phase change occurs, presents significant advantages. For instance, melting ice is a common phase change material which has a latent heat enthalpy of 335 kJ kg^{-1} and undergoes a phase transition at 0°C, and has been used in cold storage applications. One notable advantage of PCM as a cold storage medium in CES is its superior energy density and cold capacity compared to CES utilizing sensible heat materials. Furthermore, it can maintain unstable cold input and subsequently release stable cold, thereby

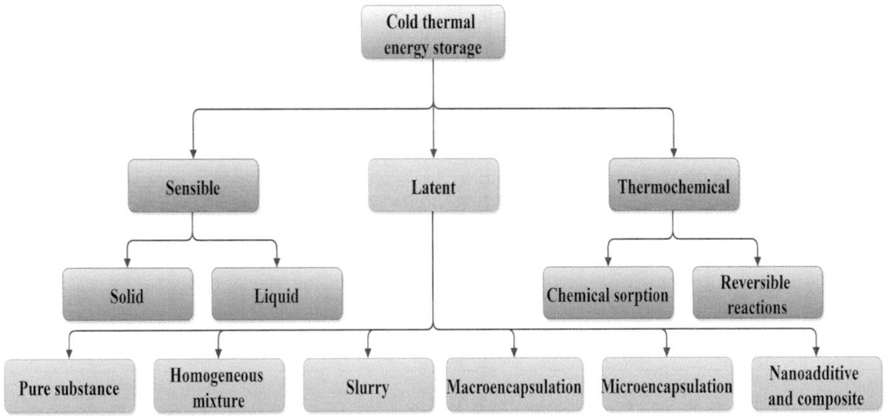

FIGURE 4.3
Classification of cold thermal energy storage.

enhancing the reliability of the cooling system due to the consistent phase change temperature [5]. Figure 4.3 represents the classifications of CES.

4.2.1.1 Sensible Thermal Energy Storage

Materials designed for thermal energy storage can retain thermal energy by changing in their temperature in the form of heat or cold. The specific heat capacity of a material is determined by its mass, allowing for the calculation of the energy stored within a sensible material using the following equation (4.1):

$$Q_s = \int_{T_{initial}}^{T_{final}} (m.c_p)\,dT \tag{4.1}$$

Where Q_s in KJ is the sensible thermal energy stored, c_p in kJ kg^{-1} K^{-1} and m in kg are the specific heat capacity of storage material and total mass, and T in K is the temperature. Sensible energy storage for CES is more direct and economical than latent storage methods for thermal energy [6]. The primary materials appropriate for CES are essentially solid in composition. The study of storage materials across a temperature spectrum ranging from room temperature down to approximately –170°C has been carried out on various substances, such as plastics, metals, rocks, concrete, quartz, and rock pebbles. Numerical and experimental investigations of the liquid air energy storage LAES system have shown that high-grade cold energy storage can be utilized efficiently through the utilization of packed beds containing rocks. Investigations have concentrated on commercial polymeric substances,

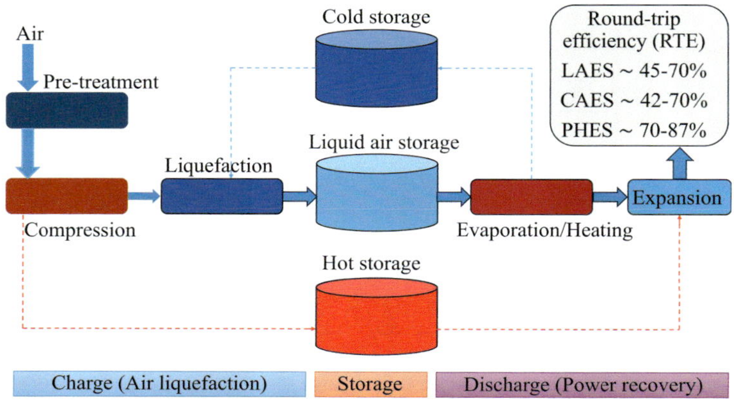

FIGURE 4.4
LAES system with cold and hot thermal recycled.

particularly polypropylene (PP) and high-density polyethylene (HDPE), as effective solid, sensible CES materials. The specific heat capacities of these polymers exhibit similarities, making them suitable for packed-bed systems. Future studies should concentrate on more robust materials, especially those sourced locally and recycled, provided they exhibit stability at extremely low temperatures and possess suitable thermophysical characteristics. Figure 4.4 provides an overview of sensible thermal energy storage.

In this context, liquid sensible thermal energy storage, as compared to solid sensible materials used in a CES system, can function simultaneously as thermal storage materials and heat transfer fluids. Nonetheless, applications must operate within a spectrum beneath these substances' freezing and boiling thresholds. A standard for selecting liquid sensible substances, as proposed by Laughlin, suggests that these materials must remain in a liquid state at lower temperatures while ensuring their vapor pressure does not exceed 1 atm [7]. Hydrocarbons and their derivatives predominantly meet the criteria for liquid sensible materials. Laughlin proposed the use of hexane as an example of a pumped energy storage system. However, health and safety concerns arise due to their potential toxicity and flammability. To enhance the combustion resistance of hydrocarbons, Laughlin has recommended techniques such as halogenation. An ethylene glycol-water solution illustrates a significant case of sensitive storage materials. The substance is defined by its freezing point under 0°C, colorless, and a reduced vapor pressure. Shorter-chain alcohols, including ethanol, propanol, methanol, propylene glycol, and their aqueous solutions, offer promising options for consideration as materials for storing liquid sensible thermal energy. Nonetheless, the most fluid and sensible substances show increased viscosity under cryogenic conditions, requiring careful selection of operating conditions. Figure 4.5 shows the schematic diagram of a liquid air energy storage system.

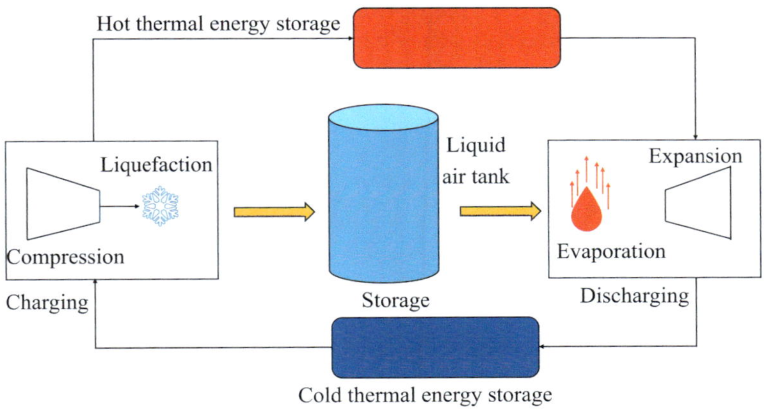

FIGURE 4.5
The schematic diagram of the liquid air energy storage system.

4.2.1.2 Latent Thermal Energy Storage

Materials designed for latent heat storage can absorb and emit thermal energy throughout the phase transition process and are called PCM. The level of temperature at which a phase transition occurs is known as the phase change temperature (PCT). The equation below can be used to determine the latent heat stored by a PCM:

$$Q_1 = x_1.m.\Delta H_1 \tag{4.2}$$

Where x_1 is the proportion of the storage material that underwent a phase change and ΔH_1 [kJ kg^{-1}] is the specific latent heat of the material. The total energy stored will include the contributions from sensible heat and latent heat. PCMs have undergone extensive development over the past several years. These materials can store significantly greater energy than sensible thermal energy storage options. However, the structure of latent thermal energy storage materials is more complex than that of sensible thermal storage materials [8].

Latent thermal energy storage encompasses four types of phase transitions: liquid-gas, solid-liquid, solid-solid, and solid-gas. The solid-liquid form is the most extensively studied and utilized category of PCM. This transformation exhibits greater latent heat compared to the solid-solid type. The thermal expansion of materials during a phase change can reach 10% of the initial volume or less. The latent heat released during the solid-solid transition is lower than that observed in the solid-liquid phenomena. Transitions between solid-gas and liquid-gas phases can store and release substantial thermal energy, which results in notable volume changes during the phase change process. This necessitates the use of storage vessels or open-air systems.

In this scenario, the only feasible options for atmospheric emission are air, nitrogen, and carbon dioxide, as they can be released without risk. The PCM products are divided into six categories: homogeneous mixture, nanoadditive and composite, microencapsulated, pure substance, macroencapsulated, and slurry PCMs.

4.2.2 Phase Change Materials

Cryogenic cold storage technology presents significant potential within the realm of refrigeration technology. PCMs are important in effectively releasing stored cold energy when subjected to elevated temperatures, which enables efficient energy transfer over time and space dimensions. Furthermore, methods for phase change cold storage have been effectively improved to extend the time involved in processes for controlling phase change temperatures, thus addressing the cost challenges linked to refrigeration transportation. Professionals in these areas have shown considerable interest in employing the phase change cold storage technology to address their cold storage requirements. A technical approach for a phase change cold storage system has been presented, including a heat discharge loop, a heat recovery loop, and PCMs to facilitate efficient heat exchange. This design prioritizes compactness and minimizes internal energy consumption without any extra energy input [9]. A cooling storage mechanism includes a refrigeration unit, a cooling storage unit, and a cooling unit. This innovation addresses the issue of inconsistent variation in temperature within the cooling storage unit and the phenomenon of phase-change ice floating. The fundamentals of phase change materials are illustrated in Figure 4.6. A device for cooling storage that utilizes phase change technology has a cooling system featuring a shell,

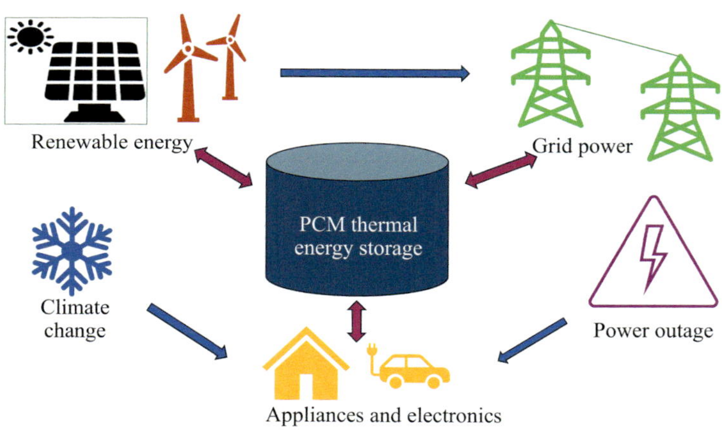

FIGURE 4.6
Fundamentals of phase change materials for energy storage.

materials for energy storage, a water inlet pipe, and an outlet pipe. Therefore, achieving a uniform distribution of water within the energy materials for storage is advantageous to tackle temperature challenges effectively, thus improving cold storage efficiency and extending the lifespan of the materials. The implementation of phase change cold storage is set to enhance energy efficiency in cold storage systems. Therefore, selecting an appropriate PCM is a crucial determination.

PCMs are categorized into four types according to their phase change mechanisms: solid-solid, solid-liquid, solid-gas, and liquid-gas. The fluctuations in volume and pressure within solid-gas and liquid-gas phase change systems pose considerable challenges, restricting their use in large-scale applications. Moreover, it is essential to acknowledge that solid-solid phase change systems exhibit low thermal conductivity and restricted phase change heat transfer capacity, reducing their potential for widespread application [10]. Consequently, solid-liquid phase change materials are mostly used. The process requires an internal molecular rearrangement that shifts from an organized state to a disordered one, enabling the absorption and release of heat energy. A solid-liquid phase change system exhibits minimal volume variation and maintains a stable temperature throughout the transition from solid to liquid, making it the most beneficial option for PCMs. Additionally, PCMs can be categorized primarily into organic, inorganic, and common crystal types according to their composition.

4.2.2.1 Organic Phase Change Materials

Alkanes represent a class of hydrocarbons that display saturated chain structures. Carbon-based hydrocarbons are primarily divided into two significant categories: alkanes and non-alkanes, especially when considering their role in organic phase transition materials. The latent heat linked to phase transitions typically shows a progressive increase as molecular weight and carbon number rise. Organic PCMs demonstrate features including phase-free separation, self-nucleation, exceptionally high latent heat of phase change, and remarkably stable chemical properties. Nonetheless, the primary drawbacks of these materials lie in their limited thermal conductivity and the gradual reduction in latent heat during the phase change process. Single-component alkanes and paraffin are the primary materials for cold storage, experiencing phase transitions in alkanes. Key characteristics encompass low corrosiveness, stable chemical properties, excellent solid-state formability, minimal supercooling, and the absence of phase separation. The main limitations include flammability, restricted energy storage density, and limited thermal conductivity. The choice of different alkanes, based on specific criteria, is thus achievable [11].

The practical benefits of non-alkane phase change cold storage materials have significantly diminished due to their inadequate thermal conductivity and associated toxicity issues. The significance of nanometal, metal-based materials, expanded graphite, graphite powder, and carbon fiber in enhancing thermal conductivity and heat transfer in organic PCMs is increasingly recognized. The phase change temperature of a binary organic phase change cold-storage material ranged from approximately 0°C to –3°C [12]. This formulation consists of tetradecane and n-octanoic acid in a specific ratio. The latent heat of phase transition is measured at 191.8 J/g, with a thermal conductivity of 0.379 W/(m K), and a minimum eutectic temperature observed at 1.0°C. Expanded graphite was integrated with n-octanoic and decanoic acid, following a mass ratio of 71:29, to create composite PCMs for pharmaceutical refrigeration and delivery systems. The findings indicate that the latent heat of the composite PCM is measured at 112.7 J/g, with a phase transition temperature recorded at 0.9°C.

4.2.2.2 Inorganic Phase Change Materials

The primary inorganic phase transition cold storage materials classifications include metal alloys, inorganic compounds, and hydrated salts. Solidification and melting in PCM are associated with water absorption and dehydration of hydrated salt. Table 4.1 presents the physical characteristics of frequently utilized inorganic phase transition materials for cold storage applications. In the processes of dehydration or melting, when crystals experience dissolution, they could lose some or potentially all crystal water. Despite their low cost and significant latent heat storage capabilities, the broader application of hydrated salt phase change cold storage materials is limited by challenges such as supercooling, phase separation, and the corrosion of metal containers [13].

The concept that the lack of carbon exclusively characterizes inorganic compounds is a widespread misunderstanding. It is essential to acknowledge that specific carbon-containing substances, such as bicarbonate, carbonate, and cyanide, are classified as inorganic compounds. Typically, the latent heat linked to phase change is relatively minimal. Ice is recognized as one of the oldest and most prevalent inorganic substances, noted for its high specific heat, considerable latent heat during phase transitions, affordability,

TABLE 4.1

Inorganic PCM Materials Used for Refrigeration

Item	Material	Chemical formula	Latent heat kJ kg^{-1}	Temperature	Ref.
1	Potassium fluoride tetrahydrate	$K_2HPO_4 \cdot 6H_2O$	109	14	[15]
2	Lithium chlorate trihydrate	$LiClO_3 \cdot 3H_2O$	253	8	[16]
3	Potassium fluoride tetrahydrate	$KF \cdot 4H_2O$	236.1	–21.2	[17]
4	Dipotassium hydrogen phosphate tetrahydrate	$K_2HPO_4 \cdot 4H_2O$	231	18.5	[18]

eco-friendly characteristics, stable phase change temperature, and improved thermal conductivity compared to water.

The behavior of various inorganic hydrated PCMs, especially their tendency for supercooling and phase separation, necessitates the incorporation of suitable thickeners and nucleating agents to enhance their properties. Identifying nucleating agents with lattice properties that closely match those of the target materials is essential for advancing future investigations. Several thickeners exhibit inadequate thermal conductivity, leading to increased phase separation phenomena and a reduction in the material's thermal performance. Consequently, upcoming investigations will identify thickeners with enhanced quality and improved thermal conductivity [14].

4.2.3 Dynamic Ice Storage

The rapid exhaustion and ineffective use of finite energy supplies have resulted in significant worldwide energy shortages, rising energy costs, and the onset of associated environmental issues, including climate change and the urban heat island phenomenon. The UN Panel on Climate Change highlights the importance of achieving net-zero carbon dioxide emissions by 2050 to fulfill the temperature targets set by the Paris Agreement. The Chinese government has pledged to attain carbon neutrality by 2060 and to lower peak CO_2 emissions by 2030 [19]. Enhancing energy efficiency and boosting the proportion of renewable energy sources in the energy mix, consumption, and emissions are essential for achieving these global and national goals. The construction sector represents approximately 40% of worldwide primary energy consumption, positioning it as the largest energy consumer on a global scale [20]. Ventilation, heating, air conditioning, and refrigeration systems constitute a substantial fraction of a building's overall energy usage, varying from 50-60%. Thus, measures focused on improving energy efficiency and maintaining HVAC and R systems have become crucial for attaining carbon peak and carbon neutrality goals.

Figure 4.7a shows the generation emissions trajectory across diverse scenarios. It shows the time-series trends in emissions from 2025 to 2050. It also shows the impacts of the diverse scenarios considered in the study. These include MidCase, LowRECcost, HighNGPrice, etc. It shows that the emissions decline significantly from 2025. However, the decline is more pronounced in the scenarios where the renewable energy costs are low (LowRECost) and natural gas prices are high (HighNGPrice). Figure 4.7b shows the system cost projections for the scenarios considered. In this case, the system costs decline and then increase. However, the increase in the system costs is scenario-specific. For instance, the system costs are lower where renewable energy costs are lower. On the other hand, the scenarios where the prices of natural gas are high show an increase in the system costs.

As illustrated in Figure 4.7c, the projected trends in the evolution of the electricity demand for the 2025~2050 period show a steady and linear increase

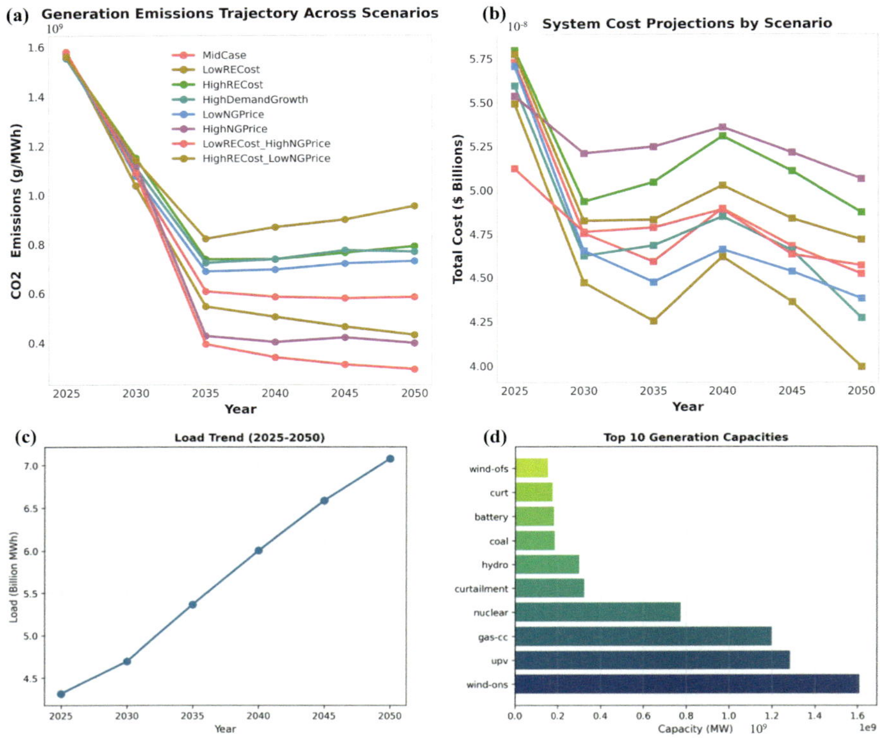

FIGURE 4.7
Comprehensive analysis for energy storage from 2025 to 2050.

in the amount of energy demanded per year, starting from 4.5×10^6 MWh and going above 7×10^6 MWh. This steady increase in the amount of energy demanded, which can be attributed to the cumulative effects of population and economic growth, and most importantly, the accelerated electrification of end-use sectors such as transportation, space heating, cooling, and industrial processes, among others, points to a trend that is not significantly influenced by short-term shocks. From a power system planning perspective, this trend suggests that there is an ongoing and ever-increasing requirement for capacity expansion, grid reinforcement, and demand management to ensure the overall reliability and adequacy of the power system while at the same time catering to the ever-increasing demand.

Figure 4.7d shows a summary of the top ten installed generation capacity types, which clearly shows that the most dominant generation capacity types are those of wind and utility-scale photovoltaic (UPV) resources, followed by nuclear and natural gas-based generation. This shows that there is a decisive move towards renewable energy resources in line with the overall push towards decarbonization and climate change mitigation. The fact that wind and UPV resources are at the heart of this generation capacity, while nuclear provides a reliable and low-carbon source of energy that improves the overall reliability and stability of the power system, while natural gas provides a source of energy that can be easily adapted to provide the much-needed flexibility in the overall energy mix, points to a trend that is highly desirable and sustainable in the long term. In addition, the fact that battery energy storage and curtailment are also among the top ten generation capacity types shows that there is an increased reliance on energy storage and its role in mitigating the intermittency of renewable energy resources, minimizing energy spillage, and improving the overall stability of the power system.

CES has considerable potential for integration with air conditioning systems by efficiently decreasing peak loads and reallocating energy consumption to off-peak periods. During off-peak periods, cold energy is produced and stored for efficient use when demand increases and electricity prices rise. The average load often stays below the previous peak, cooling demand, allowing the use of smaller chillers and consequently improving design efficiency. Furthermore, CES improves operating flexibility, enabling air conditioning systems to provide a broader spectrum of cooling requirements than systems without storage capacity. This method is successful and economical for entities like office buildings, retail stores, and educational institutions that mostly function during daylight and are closed on weekends [21].

Compared to the other two forms of CES, an ice storage system notably exhibits substantial advantages, including reduced reservoir volumes and increased cooling efficiency. Water is an energy-storage medium in ice storage by leveraging its phase-change properties to absorb the substantial latent heat of fusion acquired during the charging cycle. The ice storage process uses off-peak energy during night hours to produce ice in the storage unit, which is then retained for later utilization. The ice will gradually melt throughout the day, providing the necessary cooling capacity to match the air conditioner's installed capacity and the power grid's electrical consumption at peak hours. Accurate execution of an ice storage system could reduce energy and operational expenses for customers. With the progression of ice storage technology, an increasing number of commercial buildings are implementing ice storage systems to address the challenges posed by fluctuating energy consumption throughout the day and night. This development has led to considerable economic advantages [22].

Many specialists around worldwide have examined ice storage methodologies. In combination with photovoltaic panels, a solar-powered cooling system was developed that has an internal melted ice-on-coil storage tank. The research reveals that implementing the suggested system in an office building might result in a 38–40% decrease in energy usage and CO_2 emissions compared to traditional cooling systems. A comprehensive cold storage system was developed that effectively combines chilled water storage with seasonal heat-pipe-based ice storage, utilizing the natural cold of winter for summer-free cooling applications. The findings indicate that integrating two CES systems could enhance cold storage capacity and lead to a 40% reduction in life cycle costs through efficient optimization. A detailed rule-based control approach has been created for a chiller unit operating alongside an internal melted ice-on-pipe storage tank. The results indicate that, across all analyzed scenarios, the generalized rule-based controller operates within 20% of the optimal controller. Furthermore, in 53% of the examined instances, the actual cost may be within 10% of the ideal cost. Several management solutions related to ice storage systems have been investigated to reduce operational costs and energy usage.

An essential factor influencing the economic viability of ice storage air conditioning systems was the necessity for swift adjustments to electricity pricing frameworks. The optimal configuration of the ice storage system within the integrated energy system park can be determined by applying genetic algorithms. The algorithm helps to evaluate and enhance comprehensive energy system planning across diverse operational scenarios for ice storage. Four storage alternatives were analyzed across five high-temperature regions to assess the economic effectiveness of ice storage systems. This study aimed to evaluate the feasibility of shifting the cooling load from peak to off-peak hours as a strategy for cost reduction. The payback periods for off-peak electricity tariffs consistently show under seven years in an extremely high-temperature range. Nonetheless, most of the studies have concentrated on static ice storage techniques. They produced ice bonds to the cooling surface of a static ice storage system, resulting in a fixed ice layer [23].

Dynamic ice storage systems periodically transport ice from a refrigeration unit to an auxiliary storage tank. The standard operation utilizes the collected energy to supply chilled water to the system while simultaneously recharging the storage tank with water circulating through the ice. The ice slurry storage system is crucial in dynamic ice storage systems. This technique facilitates effective heat transmission between water and refrigerant, leading to ice slurry generation by mixing and suspending fine ice crystals within the liquid. Water is the most employed liquid, with small amounts of chemicals such as ethanol and glycol added to reduce the freezing point slightly. Ice slurry is ecologically safe, and its non-toxic properties designate it as a well-known PCM. The phase change characteristics of the material exhibit remarkable stability even after millions of charging and discharging cycles, demonstrating outstanding durability [24].

The significance of ice slurry storage systems has grown noticeably in recent years. A comparative analysis of domestic ice slurry and traditional storage systems was performed. The results demonstrate that the energy expenses linked to an ice slurry storage system are noticeably low compared to those of a conventional system, accounting for less than 50% of the overall costs. An enhanced vacuum cooling system is presented with the capacity to store ice slurry. The results demonstrate that cold storage may significantly decrease the initial and repeating expenses related to the vacuum

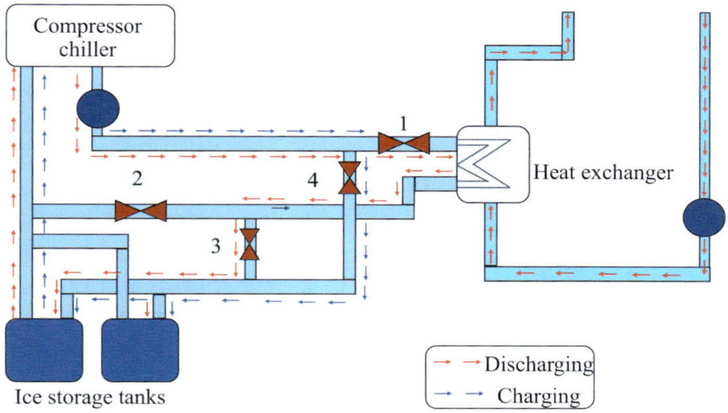

FIGURE 4.8
Schematic of dynamic ice slurry thermal storage system.

cooling system, resulting in a remarkable 29% reduction in life cycle costs [25]. Despite the potential benefits, the implementation of ice slurry storage systems in buildings with high cooling requirements remains constrained, and the evaluation conducted to analyze the system is not thorough. The illustration of the dynamic ice slurry thermal energy storage system is shown in Figure 4.8.

4.3 Applications of Cold Energy Storage

Systems that generate, use, or store cold energy at temperatures below 0°C use sub-zero thermal energy storage techniques. CES is an efficient approach to improve efficacy, reduce energy consumption, and enable power load shifting in active and passive refrigeration systems operating at sub-zero temperatures. CES fulfills a distinct role within the space exploration industry. The effective integration of renewable energy sources into traditional power grids will largely depend on developing novel electricity storage technologies. Integrating CES technology is essential for these systems, which may significantly enhance their functionality. Liquid air energy storage, pumped thermal energy storage, and superconducting flywheel energy storage are prime examples of systems that utilize CES. Although PTES and SFES remain in the prototype phase, LAES has effectively incorporated CES into commercial applications. The integration of CES with various energy systems that use off-gas cold energy offers the potential to improve energy efficiency. This cover includes high-pressure hydrogen storage tanks, re-gasification facilities, and unloading stations at an LNG port.

4.3.1 Active Refrigeration

The primary method for producing cold energy for active cooling utilizes refrigeration cycles. Numerous studies have investigated PCM-based CES for active refrigeration systems. The CES units relevant to this application are often situated in the storeroom next to the evaporator. The CES unit may be positioned between the refrigerated compartment and the refrigeration system [26]. The slurry-based CES system, designed to mimic cold absorption from the primary loop, is a crucial component of the secondary cooling system. The accumulated cold is then used to satisfy cooling demands later or in an alternate place. Figure 4.9 provides an example of using PCM panels for a freezer, and Figure 4.10 depicts a slurry system as the secondary refrigeration loop. Employing CES in sub-zero temperature ranges for active refrigeration systems has many advantages: (1) Increased system efficiency which results in energy conservation, particularly during peak demand; (2) Minimized compressor operation time; (3) Enhanced precision in temperature control; (4) Extended temperature stability during power outages; (5) Reduced operational noise; (6) Lowered greenhouse gas emissions; and (7) Decreased overall expenses for refrigeration systems.

Installing a CES unit inside a comprehensive refrigeration system optimizes the exploitation of renewable energy sources and permits effective load shifting. The implementation of CES was studied inside a hybrid solar-powered water-ammonia absorption refrigeration system. The concentrated solar dishes create 1649 kW of heat energy by capturing solar radiation. The system provides 373.7 kW of cooling electricity throughout the day, with a part employed to run the refrigeration system directly, while the remaining is kept in the CES unit at –23.51°C. During the night, the collected cold energy is consumed. The operating dynamics evaluated an industrial

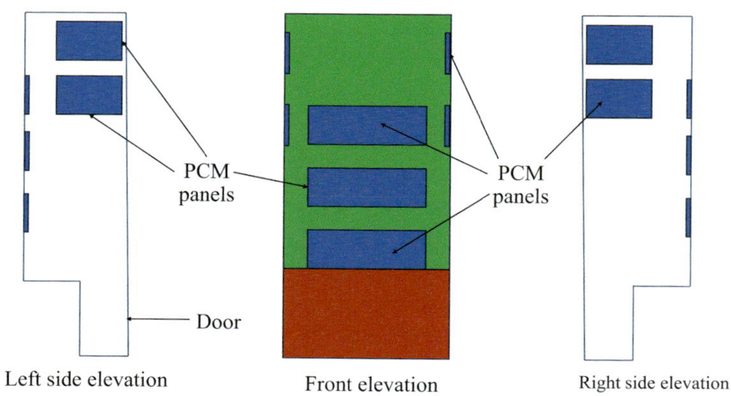

FIGURE 4.9
Example of using PCM panels for a freezer.

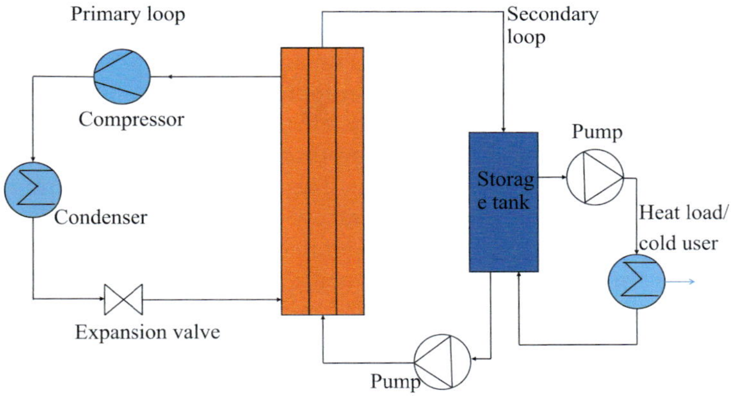

FIGURE 4.10
Slurry system as the secondary refrigeration loop.

NH_3/CO_2 cascade refrigeration system containing six large-scale CES modules inside a chicken processing plant [27]. Within the CES system, a certain PCM displays a freezing point of –11°C. The simulations demonstrated that whereas the alternative approach resulted in a small 3% drop in COP, the CES system produced a considerable 19% reduction in compressor effort during discharge.

CES systems may be used for chilling applications across diverse sectors, including space exploration and residential refrigeration. NASA formulated and evaluated the exposure sequence of CES technology for space exploration applications. Sensor components that react to temperature fluctuations and have been condensed and stabilized using PCMs include focus planes, optics, mirrors, and telescopes. This investigation centers on 2-methyl pentane (isohexane), comprising 3% acetone and a PCT of –153°C. Orbital and ground trials have shown that the technique can significantly decrease the dimensions of cooling systems used for spaceborne infrared sensors. The heating of the sensors and ambient temperature needs significant cooling. CES technology enables a more direct method for reducing variations in cooling demands associated with refrigeration. The PCMs efficiently reduce cooling requirements during operation by melting and releasing cold energy while recharging during off-peak hours. CES significantly enhances temperature management by effectively regulating PCM's operating temperature. The integration of CES could lead to a significant reduction in the size of radiators or refrigerators. CES provides weight and power savings compared to traditional refrigeration systems. Furthermore, the absence of moving parts in CES diminishes vibrations that may disrupt sensors. Refrigerators can be activated during the sensor's working hours and deactivated during the non-operational phase while CES is being charged.

4.3.2 Passive Refrigeration

A passive refrigeration system depends on an existing supply of cold energy to efficiently retain the cold, monitor temperature elevation, and extend the cold storage period. CES systems may efficiently manage passive cooling applications to transport and store temperature-sensitive products, including flowers, medicines, food, and drinks. The primary use of big microencapsulated CES is in transporting temperature-sensitive products. Additional CES units may be integrated into refrigerated vehicles. The benefits of pure PCM systems are assessed in food delivery. CES plates or beams may be used independently or with traditional vapor compression systems in refrigerated vehicles. Effectively preloading CES units overnight and using passive solar cooling to maintain a suitable temperature inside the car throughout the day. The environmental effect of refrigeration in diesel food trucks is around 40%, signifying a considerable possibility for improvement using CES passive cooling technology. The CES devices exhibit significantly improved efficiency when energized by renewable energy sources. Commercial storage operations may benefit from using CES systems in sub-zero temperature settings. Viking Cold Solutions, Inc. claims that its passive PCM cooling system may reduce refrigeration expenses for warehouses operating at –18°C by about 40%.

4.3.3 Electricity Storage

Incorporating a larger share of renewable energy sources, particularly solar and wind, is crucial for enabling future energy transitions and alleviating the effects of climate change. However, there exists a degree of inconsistency among these sources. Energy storage technologies are essential for reducing peak-load capacity and facilitating the equilibrium of the grid's real-time demand-supply dynamics. Promoting the integration of renewable energy sources into power grids facilitates the improvement in the demand and supply chain. Various concepts for energy storage systems have been developed or built, with CES playing a crucial role in some of the most innovative and promising designs. LAES, PTES, and SFES represent unique technologies that utilize or can implement CES [28].

4.3.3.1 Liquid Air Energy Storage

LAES technology employs liquefied air at –196°C to store and release energy in electricity [29]. The standard LAES operates via three phases: charging, storing, and recovering energy. In Step 1 of charging, the LAES system accumulates supplementary electrical energy, often more economical during off-peak hours, to purify, compress, and liquefy the air. Step 2 requires the storage of liquid air within inertially insulated tanks, maintained at a temperature of –196°C and close to ambient pressure. Step 3 entails discharging during energy recovery while pumping, reheating, and expanding in response to elevated demand and increasing prices to produce power during peak hours when electricity costs are high. This step requires simultaneous pumping and heating of liquefied air alongside the discharge of the succeeding phase. The energy acquired from the reheating process is preserved in traversing the pumping system and the heat pump. Steps 1 and 3 may decrease liquefaction expenses and improve electricity generation using stored thermal energy. The cold energy emitted during the discharge phase of the LAES system is retained in a packed-bed type CES inside the first pilot plant. The amount of cold energy available for charging is a vital factor in the overall performance metrics of the LAES system. The round-trip efficiency increased by 20%, and liquid output improved by 30%, owing to the integration of 16% recovered cold energy.

4.3.3.2 Pumped Thermal Energy Storage

Pumped thermal energy storage, often known as pumped thermal electricity storage, is an alternative approach for large-scale energy storage. The CES is the primary component of the PTES. Thus far, the PTES technology and CES system have mostly been in the initial stages of development, concentrating largely on system design and the installation of pilot plants. Isentropic Ltd. and Newcastle University have partnered to develop a prototype facility with 600 kWh storage capacity [30]. The two thermal energy storage facilities comprising a PTES, one considered for hot and the other for cold TES, are interconnected by a heat pump or engine cycle. The diagram demonstrates the functionality of the PTES system and presents the setup during the charging process with essential outlines. The system now functions as a heat pump, using surplus power to transfer heat from the cold storage to the hot storage. During the discharge phase, the heated fluid from the hot storage cools, produces electricity via an expander, and then returns to the cold storage, operating somewhat like a heat engine. The layout of the pumped thermal energy storage system is depicted in Figure 4.11.

Extensive studies on latent heat CES using salt-water mixtures as working fluids have been recorded; nevertheless, it seems that most studies have focused on sensible CES for broader applications associated with PTES. The operational parameters suggest that the CES facility must be equipped to store cold energy at temperatures of about 100 K. Thus, conventional sensible heat thermal energy storage systems would have homogeneous ceramic matrices or tightly packed aggregates of gravel or pebbles. The maximum density of gravel is 50 kWh/m^3, whereas compressed air and pumped hydro energy storage need 10 kWh/m^3 and 1.4 kWh/m^3, respectively. A liquid thermocline would exhibit significantly enhanced efficacy in reducing energy loss relative to a solid-packed bed. The suggested division between the warm and cold zones in the liquid thermocline aims to reduce supplementary energy losses. According to certain assumptions, N-hexane is considered an

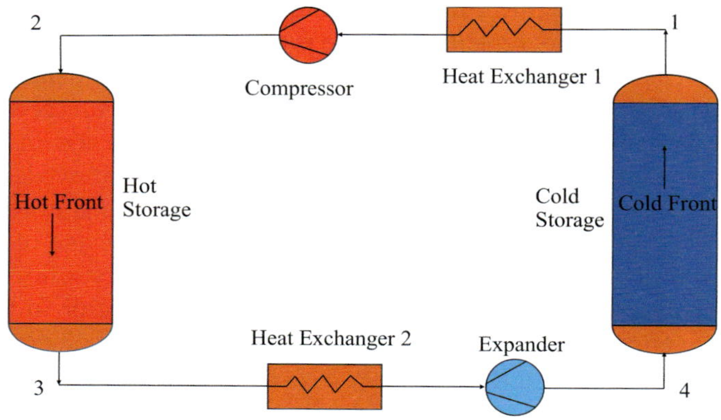

FIGURE 4.11
The layout of a pumped thermal energy storage system.

exceptionally efficient liquid cold storage medium for PTES. The flammability and neurotoxicity of this material pose considerable issues, notwithstanding its acknowledged cost-effectiveness and biodegradability.

4.3.4 Waste Cold Energy Recovery

Various energy processes, such as LNG regasification and the release from high-pressure hydrogen storage tanks, result in considerable cold energy loss. The integration of CES with diverse energy systems may improve overall efficiency, diminish energy consumption, and stabilize variations in energy supply. LNG has become an essential component of the energy portfolio for many nations worldwide, especially in Europe and Asia. Proposals have been put out for devices designed to recover the substantial LNG-cold energy losses during the regasification process. The demand for LNG varies daily and seasonally, shaped by the changing needs for city gas and electricity, complicating the simple use of cold energy. Thus, CES can alleviate the variations associated with cold energy supply and enhance the efficient use of LNG cold energy. Despite its considerable potential, the use of CES for LNG cold recovery is still very limited [18].

During daylight operations with substantial LNG flow, the boil-off gas (BOG) may be re-liquefied at the specified pressure by using the cold energy from the LNG. The CES employs a phase transition material to accumulate surplus cold energy, reaching a freezing point of about –130°C. The LNG flow alone may be inadequate for the LNG/BOG two-phase flow to efficiently re-liquefy the BOG during overnight hours. In these cases, BOG is re-liquefied using the stored cold energy from the CES. Research on CES trials suggests that using the CES method for re-liquefaction, rather than traditional BOG compression, may reduce compression power by 30–60%. An innovative concept was introduced about the use of PCM for the storage of liquefied natural gas (LNG) cold energy for potential applications. Although carbon dioxide and air are efficient for power production, PCMs may be used to improve the cooling of distribution systems. Investigating supplementary CES systems that use CES for LNG cold energy utilization is essential.

4.3.5 Applications of Artificial Intelligence for Cold Energy Storage

Artificial intelligence is regarded as devices programmed with human capabilities, including problem-solving and learning capacities. The decisions by AI and validations are always optimal to ease access to certain goals. Implementing AI does involve many benefits, as it decreases human errors and speeds up decision-making. Many problems are solved using AI, compared to those requiring much effort and time in conventional computing, which can solve only one problem. Weak artificial intelligence is a software of machines, which responds with a limited, specific response according to rules set up by the user and remains within the bounds of such rules [31]. Strong artificial intelligence comprises the entire subset of machine learning and deep learning. Machine learning algorithms are made to learn and improve accuracy through the given effective data input-output. Most of the articles published in recent years have discussed artificial intelligence implementations in diverse areas of engineering, cybersecurity, medicine, arts, and transport. However, AI can generate countless features for most complicated energy issues or renewable energy in general: (1) design and modeling of different energy systems, (2) forecasting energy generation and demand, (3) maximizing overall system performance by minimizing energy cost, and (4) management of energy between diversified components of the energy system. Several artificial intelligence techniques and methods have emerged and been used in the various fields of the energy sector today, such as energy system design, energy efficiency monitoring, energy generation forecasting, and energy storage [32].

Figure 4.12a illustrates the evolution of the share of artificial intelligence in patent activity in various sectors from 2013 to 2022. Medical devices and imaging show the strongest and most consistent growth, with AI patent activity rising from below 1% in 2013 to a peak of almost 5% in 2020–2021 and then falling in 2022. In contrast, clean energy patent activity shows a steady and gradual rise, remaining below 1% throughout the period, indicating a slower development of AI in this sector. Finally, fossil fuels show a peak in AI patent activity in 2020, followed by a decline in 2022, indicating a less consistent development of AI in this sector compared to the strong and steady rise in the case of medical devices and imaging.

Figure 4.12b examines the share of AI in venture capital funding in various sectors from 2013 to 2024 and shows a very different trend in the evolution of AI patent activity in the same sectors. Digital sectors show the strongest growth in AI-related venture capital funding, with a sharp rise in funding

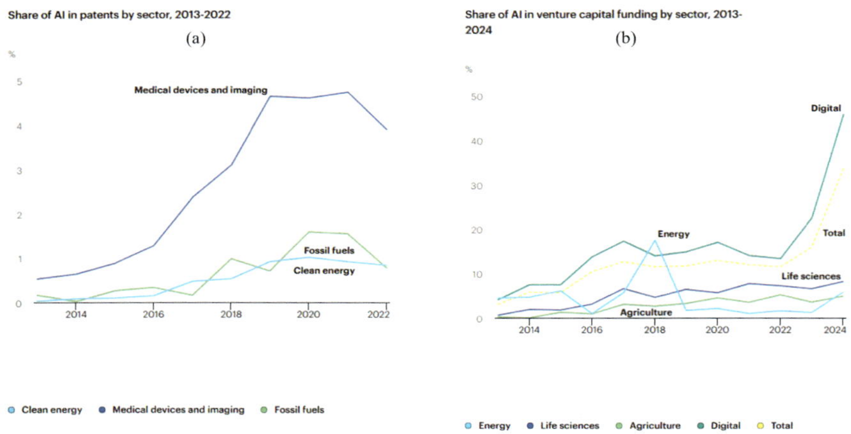

FIGURE 4.12
Applications of artificial intelligence in different fields.

share in 2022 and 2024, reaching almost half of the total AI venture capital funding in 2024. Life sciences show a steady rise in AI patent activity with moderate fluctuations, indicating a sustained investor interest in AI applications in this sector. In contrast, the share of AI in venture capital funding in the energy and agriculture sectors is much smaller, with a short-lived peak in the case of the energy sector in 2018, which does not sustain over the entire period.

The major transition from conventional energy sources into renewables and energy storage is accompanied by models, tech upgrades, and services promising evolution. Intelligent energy systems connected with artificial intelligence are now being introduced, wherein the infrastructure links the system's representations with real-life. Currently, it focuses on energy storage under AI techniques to improve energy forecasts, predict the operation of different system parts, and evaluate and analyze the system's performance. Ground-breaking advances were made possible due to technology specified to control the TES systems in forecasting or optimization. System modeling involving PCM using classical physics-based approaches proves difficult and consequently needs complicated mathematics for every component of such a system. On the contrary, using an Artificial Neural Network model simplifies many problems by considering internal processes via the direct generation of models based on input and output datasets from a physical system without requiring detailed mathematics. The use of artificial intelligence techniques in thermal energy storage is illustrated in Figure 4.13.

The advantages of ANNs are flexibility, cost, and reliability compared to conventional data handling methods, such as CFD. An ANN is a robust and efficient tool for analyzing several heat transfer processes. It has been shown

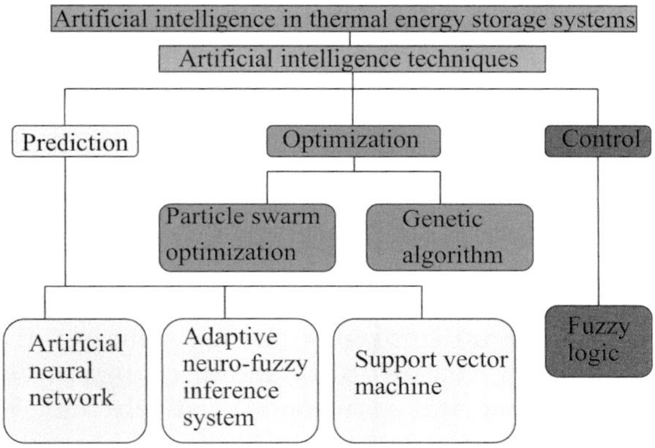

FIGURE 4.13
The use of artificial intelligence techniques in thermal energy storage.

that predictive control with an ANN could strongly control unmanageable cause-and-effect relationships with the thermal load fluctuations. It can save energy in TES water tanks by 6.5%. The fuzzy logic control kept and tracked the input signals quite well, achieving an extremely low overshoot of around 4.4% while holding the heat transfer fluid temperature to the desired one. Including FL controllers into ANN has substantially improved the performance of hybrid TESS by at least 40% regarding the reduction in costs for space heating. Although the different TES systems possess various operational or controlling parameters, the modeling and optimization of different TESS rely heavily upon generic parameters, including environmental temperature, system inlet temperature, solar radiation, wind speed, water volumetric flow rate, relative humidity, dry-bulb temperature, and ventilation rate. Modern AI techniques such as GA, PSO, and ANN simultaneously optimize multiple variables capable of predicting the efficient thermal performance of parabolic trough solar collectors in water preheating to about a 10% cost saving when integrated with GA. The use of AI techniques in various TESS has improved the overall performance of the developed systems regarding space, reduction in costs, and enhancement in efficiency.

4.4 Conclusions

The development and optimization of cold energy storage systems are considered a vital path towards the achievement of efficient energy cooling systems, grid stability, and the integration of renewable resources. This chapter provides a comprehensive overview of various cold energy storage technologies and their evaluation in terms of their storage density and efficiency. The study highlights the great potential of cold energy storage systems in addressing peak demand, carbon emissions, and cost savings in various sectors.

1. The density of cold energy storage systems is, in the final analysis, a function of the type of storage system used. In the case of sensible storage, such as the example of chilled water, the specific heat capacity is relatively low, measured at 4.18 $kJ \cdot kg^{-1} \cdot K^{-1}$. In the case of latent heat storage, which utilizes phase change materials, the density is considerably higher, thanks to the phase change enthalpy. In the case of ice, the latent heat of fusion is relatively high, measured at 335 $kJ \cdot kg^{-1}$, compared to the range of 110–250 $kJ \cdot kg^{-1}$ for organic and inorganic phase change materials.
2. In the various thermal energy storage technologies available, the most promising and fastest deployable method with cost-effectiveness and a wide range of applications is the use of chilled water storage. With regards to building suitability, however, ice storage technologies are the most promising and can save up to 38–40% of both energy and CO_2 emissions, as well as providing savings of up to 40% on lifecycle costs. To ensure minimal costs during operation, dynamic ice slurry storage technologies are the best and can reduce costs to below 50% of those of conventional technologies. The best material for thermal energy storage is ice/water due to its best properties of this material.
3. The application of Artificial Intelligence, focusing on artificial neural networks and hybrid ANN-Fuzzy systems, has proved to be quite effective for modeling and optimizing combined energy systems. The results show high accuracy while keeping error levels low, such as the maximum control overshoot being restricted to 4.4%. The results of applying AI can lead to a direct savings of 6.5% in thermal storage systems, as well as quicker computation and reliability compared to conventional physics-based modeling approaches such as Computational Fluid Dynamics (CFD).

In conclusion, the development of cold energy storage is a function of the careful selection of high-density latent storage media such as ice, as well as the development of intelligent AI-based control strategies. A comparative analysis indicates that, although sensible storage is simple to implement, the performance of ice-based latent storage provides the most favorable trade-off between performance, cost savings, and environmental benefits for the majority of building applications. The integration of these systems is critical to the development of resilient, efficient, and sustainable cooling systems of the future.

Acknowledgments

This research is funded by the National Natural Science Foundation of China (Grant No. 52576006), Beijing Municipal Commission of Science and Technology, Zhongguancun Science and Technology Park Management Committee (Grant No. Z231100006123010), and the Department of Science and Technology of Hebei Province (Grant No. 244A7625D and 254Z4504G).

References

1. Sonnenrein, G., Baumhögger, E., Elsner, A., Morbach, A., Neukötter, M., Paul, A., & Vrabec, J. (2020). Improving the performance of household refrigerating appliances through the integration of phase change materials in the context of the new global refrigerator standard IEC 62552:2015. International Journal of Refrigeration, 119, 448–456. https://doi.org/10.1016/j.ijrefrig.2020.07.025
2. Li, W. Q., Zhang, T. Y., Li, B. B., Xue, Z. R., Wang, H., & Zhang, D. (2022). Enhanced energy management performances of passive cooling, heat storage and thermoelectric generator by using phase change material saturated in metal foam. International Journal of Thermal Sciences, 184, 107869. https://doi.org/10.1016/j.ijthermalsci.2022.107869
3. Alzuwaid, F., Ge, Y. T., Tassou, S. A., Raeisi, A., & Gowreesunker, L. (2014). The novel use of phase change materials in a refrigerated display cabinet: An experimental investigation. Applied Thermal Engineering, 75, 770–778. https://doi.org/10.1016/j.applthermaleng.2014.10.028
4. Wang, G., Tang, Z., Gao, Y., Liu, P., Li, Y., Li, A., & Chen, X. (2023). Phase change thermal storage materials for interdisciplinary applications. Chemical Reviews, 123(11), 6953–7024. https://doi.org/10.1021/acs.chemrev.2c00572
5. Faramarzi, S., Mohammad Mehdi Asadzadeh, S., Reza Mohammad Hassani, M., Abdollahi, S. A., Zarehzadeh, R., & Mafi, M. (2024). Energy and exergy analysis of a modified air handling unit assisted by phase change material, heat recovery unit, and solar energy. Scientific Reports, 14(1). https://doi.org/10.1038/s41598-024-83458-5
6. Navarro, M. E., Martínez, M., Gil, A., Fernández, A. I., Cabeza, L. F., Olives, R., & Py, X. (2012). Selection and characterization of recycled materials for sensible thermal energy storage. Solar Energy Materials and Solar Cells, 107(107), 131–135. https://doi.org/10.1016/j.solmat.2012.07.032
7. Laughlin, R. B. (2017). Pumped thermal grid storage with heat exchange. Journal of Renewable and Sustainable Energy, 9(4), 044103. https://doi.org/10.1063/1.4994054
8. Sarbu, I., & Sebarchievici, C. (2018). A comprehensive review of thermal energy storage. Sustainability, 10(1), 191. https://doi.org/10.3390/su10010191
9. Zhai, X., Xu, Z., Zhang, W., Zhang, Q., Yang, X., Qu, J., Liu, G., & Yu, B. (2025). Phase change thermal energy storage: Materials and heat transfer enhancement

methods. Journal of Energy Storage, 123, 116778. https://doi.org/10.1016/j.est.2025.116778
10. Nazir, H., Batool, M., Bolivar Osorio, F. J., Isaza-Ruiz, M., Xu, X., Vignarooban, K., Phelan, P., Inamuddin, I., & Kannan, A. M. (2018). Recent developments in phase change materials for energy storage applications: A review. International Journal of Heat and Mass Transfer, 129, 491–523. https://doi.org/10.1016/j.ijheatmasstransfer.2018.09.126
11. Glova, A. D., Nazarychev, V. M., Larin, S. V., Gurtovenko, A. A., & Lyulin, S. V. (2023). Size matters: Asphaltenes with enlarged aromatic cores promote heat transfer in organic phase-change materials. Physical Chemistry Chemical Physics, 25(46), 32196–32207. https://doi.org/10.1039/d3cp02953k
12. Yadav, A., Pandey, A., Samykano, M., Kalidasan, B., & Said, Z. (2024). A review of organic phase change materials and their adaptation for thermal energy storage. International Materials Reviews, 69(7–8), 380–446. https://doi.org/10.1177/09506608241292406
13. Ismail, M., Zahra, W. K., Ookawara, S., & Hassan, H. (2023). Enhancing the air conditioning unit performance via energy storage of different inorganic phase change materials with hybrid nanoparticles. JOM, 75(3), 739–753. https://doi.org/10.1007/s11837-022-05629-x
14. Malazi, M. T., Afshari, F., & Kaya, K. (2025). Phase change materials (PCMS) for buildings and automotive applications: A review study. Heat Transfer Research, 56(1), 39–52. https://doi.org/10.1615/heattransres.2024053911
15. Veerakumar, C., & Sreekumar, A. (2016). Phase change material based cold thermal energy storage: Materials, techniques and applications – A review. International Journal of Refrigeration, 67, 271–289. https://doi.org/10.1016/j.ijrefrig.2015.12.005
16. Zhao, Y., Zhang, X., Xu, X., & Zhang, S. (2020). Research progress of phase change cold storage materials used in cold chain transportation and their different cold storage packaging structures. Journal of Molecular Liquids, 319, 114360. https://doi.org/10.1016/j.molliq.2020.114360
17. Su, W., Darkwa, J., & Kokogiannakis, G. (2015). Review of solid–liquid phase change materials and their encapsulation technologies. Renewable and Sustainable Energy Reviews, 48, 373–391. https://doi.org/10.1016/j.rser.2015.04.044
18. Fan, Y., Wang, J., Li, Y., & Hu, T. (2025). The utilization of LNG cold energy: Current status, challenges, and future prospects. Frontiers in Sustainable Development, 5(2), 7–12. https://doi.org/10.54691/8bjjd733
19. Liang, Y., Pan, Y., Yuan, X., Yang, Y., Fu, L., Li, J., Sun, T., Huang, Z., & Kosonen, R. (2022). Assessment of operational carbon emission reduction of energy conservation measures for commercial buildings: Model development. Energy and Buildings, 268, 112189. https://doi.org/10.1016/j.enbuild.2022.112189
20. Yuan, X., Chen, Z., Liang, Y., Pan, Y., Jokisalo, J., & Kosonen, R. (2021). Heating energy-saving potentials in HVAC system of swimming halls: A review. Building and Environment, 205, 108189. https://doi.org/10.1016/j.buildenv.2021.108189
21. Numminen, S., Ruggiero, S., & Jalas, M. (2022). Locked in flat tariffs? An analysis of electricity retailers' dynamic price offerings and attitudes to consumer engagement in demand response. Applied Energy, 326, 120002. https://doi.org/10.1016/j.apenergy.2022.120002

22. Erdemir, D., Altuntop, N., & Çengel, Y. A. (2021). Experimental investigation on the effect of ice storage system on electricity consumption cost for a hypermarket. Energy and Buildings, 251, 111368. https://doi.org/10.1016/j.enbuild.2021.111368
23. Rahgozar, S., Dehghan, M., Pourrajabian, A., & Haghgou, H. (2021). Economic feasibility of ice storage systems for office building applications: A climate sensitivity analysis. Journal of Energy Storage, 45, 103712. https://doi.org/10.1016/j.est.2021.103712
24. Hamid, M., Mengjie, S., Hadibi, T., Mohasan, M., Uglanov, D. A., & Zekang, Z. (2025). Research progress and challenges in the preparation and application of ice-making technology. Renewable and Sustainable Energy Reviews, 223, 115996. https://doi.org/10.1016/j.rser.2025.115996
25. Liu, S., Li, H., Song, M., Dai, B., & Sun, Z. (2017). Impacts on the solidification of water on plate surface for cold energy storage using ice slurry. Applied Energy, 227, 284–293. https://doi.org/10.1016/j.apenergy.2017.08.012
26. Zhou, S., Song, M., Shan, K., Razaqpur, A. G., & Huang, J. J. (2024). Parametric and optimization analyses of a dynamic trombe wall incorporating PCM to save heating energy under cold climate zones. Renewable Energy, 237, 121537. https://doi.org/10.1016/j.renene.2024.121537
27. Wang, H., Xie, B., & Li, C. (2024). Review on operation control of cold thermal energy storage in cooling systems. Energy and Built Environment, 6(3), 509–523. https://doi.org/10.1016/j.enbenv.2024.01.007
28. Psarros, G., Dratsas, P., & Papathanassiou, S. (2024). A comprehensive review of electricity storage applications in island systems. https://doi.org/10.48550/arxiv.2401.14712
29. Sciacovelli, A., Vecchi, A., & Ding, Y. (2017). Liquid air energy storage (LAES) with packed bed cold thermal storage – From component to system level performance through dynamic modelling. Applied Energy, 190, 84–98. https://doi.org/10.1016/j.apenergy.2016.12.118
30. Davenne, T. R., & Peters, B. M. (2020). An analysis of pumped thermal energy storage with de-coupled thermal stores. Frontiers in Energy Research, 8. https://doi.org/10.3389/fenrg.2020.00160
31. Makarov, M. Y. (2020). The impact of artificial intelligence on productivity. Economics and Management, 26(5), 479–486. https://doi.org/10.35854/1998-1627-2020-5-479-486
32. Feng, P., Jiang, G., Li, K., Xu, J., Shen, Y., Xu, W., & Wang, S. (2024). Two-stage stochastic robust optimization scheduling of electric–thermal microgrid with solid electric thermal storage. AIP Advances, 14(1). https://doi.org/10.1063/5.0185649

5

Fundamentals of AI in Energy Storage Applications

Ayesha Arif, Ameer Hamza Tiwana, and Yabin Jin

5.1 Introduction

Artificial Intelligence (AI) is revolutionizing energy systems by facilitating intelligent decision-making, predictive analytics, and optimization in real time. As renewable energy sources like solar and wind gain prominence day by day, their intermittent nature demands advanced solutions for effective storage and distribution. AI methodologies, including machine learning (ML) and deep learning, play a vital role in meeting these challenges by predicting energy generation, optimizing storage processes, and improving grid stability. Moreover, from intelligent battery management to demand response strategies, AI systems are reshaping how energy is stored, distributed, and utilized, thus fostering a more resilient and sustainable energy framework.

The integration of AI in renewable energy storage is essential for enhancing efficiency, ensuring reliability, and lowering costs. Unlike traditional power systems, where renewable energy depends on fluctuating sources, making storage optimization a complex issue. AI resolves such challenges by analyzing extensive datasets for forecasting energy supply and demand imbalances, optimizing charge and discharge cycles, and eventually extending battery life. AI also improves thermal energy storage (TES) by enhancing thermal load forecasting and energy dispatch efficiency. With AI, the energy storage systems can dynamically adjust to the fluctuations and promote the adoption of renewable energy, which eventually expedites the transition to a decarbonized grid.

5.2 Fundamentals of Artificial Intelligence (AI) for Energy Storage

5.2.1 Basic Concepts of AI and Machine Learning (ML)

Involvement of computational methods in AI based structure enable systems to learn from data by identifying patterns and making decisions with

DOI: 10.1201/9781003516132-5

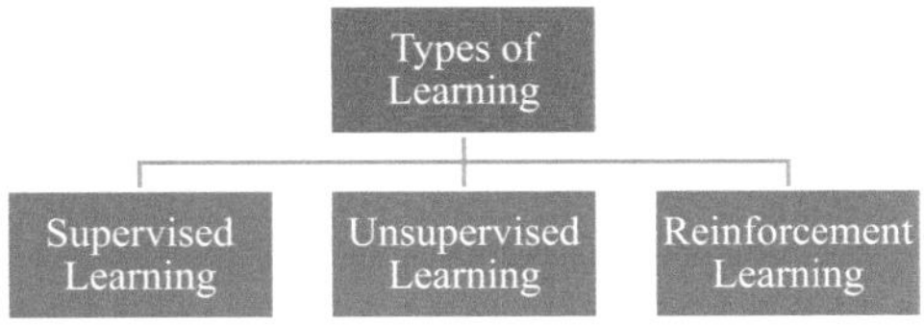

FIGURE 5.1
Types of learning. (*Source*: Authors.)

less human involvement. Machine Learning (ML) techniques are valuable in energy storage applications due to their ability to process the complex datasets and optimizing system performance. ML approaches can be classified into three learning techniques as shown in Figure 5.1:

i. **Supervised Learning:**
It uses labeled datasets for training models for prediction or classification (e.g., forecasting energy demand based on historical data).
ii. **Unsupervised Learning:**
It identifies hidden patterns in unlabeled data (e.g., clustering similar energy consumption profiles for demand side management).
iii. **Reinforcement Learning (RL):**
It trains the models through trial and error with adjustments in optimal actions (e.g., Battery charge and discharge cycles optimization for achieving maximum efficiency).

These algorithms form the basis for AI-based optimization in energy storage, which enables the management of integration of renewable energy smartly (Table 5.1).

5.2.2 Deep Learning and Its Relevance in Energy Storage

Deep Learning (DL) is a special type of ML which is based on multiple layers of artificial neural networks that model the complex patterns/relationships

TABLE 5.1
AI Algorithms Used in Energy Storage

Algorithm	Application in Energy Storage	Advantages
Neural Networks	State-of-charge (SOC) estimation, load forecasting	High accuracy, handles nonlinear data
Decision Trees	Fault detection in batteries, rule-based energy dispatch	Interpretable, fast decision-making
Support Vector Machines (SVM)	Thermal load prediction, battery degradation modeling	Effective in high-dimensional spaces

in large datasets. In energy storage systems, these deep learning techniques are used for the following purposes:

i. **Time-Series Forecasting:** Recurrent Neural Networks (RNNs) and Long Short-Term Memory (LSTM) networks enhance accuracy in predicting renewable energy generation and demand fluctuations.
ii. **Image & Signal Processing:** Convolutional Neural Networks (CNNs) analyze thermal images for fault detection in battery systems.
iii. **Adaptive Control:** Deep Reinforcement Learning (DRL) optimizes the real-time energy dispatch in hybrid storage systems.

The above-mentioned techniques have the capability to process complex data structures/patterns in energy storage and thus make it vital for the advanced energy storage solutions.

5.2.2.1 Convolutional Neural Networks (CNNs) and Recurrent Neural Networks (RNNs)

Convolutional Neural Networks (CNNs) and Recurrent Neural Networks (RNNs) are deep learning architectures and play a crucial role in energy storage applications because of their ability to process complex, structured, and high-dimensional data (Table 5.2).

- **CNNs have the ability to perform analysis of spatial patterns, making them ideal for:**
  i. Thermal image analysis (e.g., detecting battery overheating)
  ii. Fault identification in energy storage components
  iii. State of health (SOH) monitoring using sensor data
- **RNNs, particularly Long Short-Term Memory (LSTM) have the ability to process temporal data, making them ideal for:**
  i. Energy demand and generation forecasting
  ii. Predictive maintenance by analyzing time series degradation trends
  iii. Dynamic optimization of charge/discharge cycles

TABLE 5.2

Comparison of CNN and RNN in Energy Storage Applications

Feature	CNN	RNN (LSTM/GRU)
Data Type	Grid-based (images, spatial data)	Sequential (time series, sensor logs)
Key Strength	Feature extraction from visuals	Long-term dependency modeling
Energy Storage Use Case	Battery thermal imaging	Load forecasting & SOC prediction

5.2.2.2 Applications in Recognizing Pattern and Modeling Prediction

Deep learning techniques enhance the effectiveness of energy storage systems through pattern recognition and Predictive modeling, which resultantly improves their efficiency, life, and safety.

- **Pattern Recognition**
  i. **Anomaly Detection:** Abnormal battery behavior (e.g., voltage spikes) is identified by Autoencoders.
  ii. **Optimal Charging Patterns:** Reinforcement Learning (RL) combined with RNNs learns adaptive charging strategies.
- **Predictive Modeling**
  i. **Battery Degradation Prediction:** Historical cycling data for estimating remaining useful life (RUL) is analyzed by LSTMs.
  ii. **Renewable Energy Forecasting:** CNNs process weather maps, and RNNs predict solar/wind generation.

5.3 Predictive Analytics for Energy Demand and Supply

5.3.1 Time-Series Forecasting for Renewable Energy Generation

Grid integration with renewable energy sources is challenged by their intermittent nature; however, AI-powered time series forecasting can help in this regard by making accurate predictions about future energy generation by analyzing the power output data, previous weather trends, and seasonal patterns. When analyzing sequential data, sophisticated deep learning models such as Transformer topologies and Long Short-Term Memory (LSTM) networks perform exceptionally well, capturing both short-term oscillations and long-term dependencies that conventional statistical techniques frequently overlook. Grid operators can optimize the energy storage dispatch plans with the help of these AI techniques, which, on implementation, can reduce dependency on fossil fuel-based backups by up to 30%. These methods are very helpful for solar farms, which use AI for predicting changes in cloud cover and modifying battery charging cycles for optimizing the use of available sunshine.

5.3.2 AI-Driven Load Prediction for Grid Stability

Accurate load forecasting is necessary for modern power networks' stability with the increasing penetration of renewable energy sources. To forecast energy demand, AI models analyze complex and large datasets, such as past trends of consumption, current weather conditions information, and seasonal patterns. Advanced deep learning models like Long Short-Term Memory

TABLE 5.3
AI Forecasting Techniques for Energy Applications

Technique	Best For	Accuracy	Computational Cost	Example Use Case
LSTM Networks	Long-term dependencies	High	Moderate	Seasonal solar generation forecasts
Transformer Models	Multivariate analysis	Very High	High	Grid-wide demand prediction
Gradient Boosting	Feature importance	Medium-High	Low	Localized load forecasting
ARIMA Models	Short-term trends	Medium	Very Low	Baseline comparison models
Hybrid Models	Complex systems	Highest	Very High	Integrated renewable-storage systems

(LSTM) networks and Transformer architecture process the sequential data and capture short-term fluctuations and long-term dependencies, which are usually missed while using traditional statistical methods. Therefore, by enabling the dynamic energy storage management system, these AI technologies give grid operators the ability to address the possible imbalances/ outages before they occur. For example, AI optimizes the battery discharge timing to sync perfectly with the declining solar production by predicting the characteristic "duck curve" when there is a surge in evening demand.

The advanced forecasting techniques mentioned in Table 5.3, when combined with powerful optimization algorithms, create a powerful framework for managing the production dynamics of renewable energy systems.

5.3.3 AI in Battery Management Systems (BMS)

5.3.3.1 Estimation of State of Charge (SOC) and State of Health (SOH)

Accurate estimation of State of Charge (SOC) and State of Health (SOH) is one of the most crucial applications of artificial intelligence in modern management systems of batteries. The estimation of SOC involves finding out the remaining usable capacity in a battery at any given moment, while SOH determines the battery's degradation over time. Usually, the traditional methods like coulomb counting and estimation on the basis of voltage prove inadequate and inefficient due to their inability to understand the complex factors like aging, effects of temperature change, and variations in charge/discharge rate. In contrast, the modern AI approaches of machine learning algorithms such as support vector regression (SVR) and deep neural networks (DNNs) can process multiple input parameters, including temperature, voltage, current, and impedance, to deliver SOC estimates with less than 1% error in real-world conditions [1]. Similarly, for predicting SOH, recurrent neural networks (RNNs) give impressive results by analyzing the long-term cycling data for identifying complex degradation patterns that cause capacity fade. Moreover, these AI models can also predict the remaining useful life (RUL) with an accuracy of 95% approximately along with several charge cycles in advance, which resultantly enables predictive maintenance and replacement strategies [2]. Recent research also focuses on physics-informed neural networks, which combine electrochemical battery models with data-driven learning for achieving high accuracy in both SOC and SOH estimation for different batteries (i.e., lithium-ion to emerging solid-state designs) with different operating conditions.

5.3.3.2 Charge/Discharge Cycles Optimization for Battery Lifespan

AI-integrated optimization of charge/discharge cycles is a powerful technique for extending the lifespan of the battery, along with maintaining performance. Widely used Lithium-ion batteries in modern energy storage applications experience degradation due to different causes like Lithium plating, growth of solid electrolyte interface (SEI), and active material loss. However, the machine learning algorithms can analyze these thousands of charge/discharge cycles for identifying patterns that result in lower degradation effects along with meeting energy delivery requirements. Similarly, in reinforcement learning (RL) approaches, AI models learn the optimal charging strategies by continuous interaction with battery systems. For instance, the comparison of conventional constant current/constant voltage (CC/CV) methods with the "Deep Q Network" (DQN) shows that DQN can adjust charging currents dynamically on the basis of real-time conditions, which reduces the degradation by approximately up to 30% [3]. Moreover, the advanced implementations also combine model predictive control (MPC) with RL for the optimization of both longer degradation and outstanding performance simultaneously. In addition, these AI systems also consider multiple variables such as state-of-charge (SOC), temperature, and past trends of consumption for finding the ideal charging profile for different situations. Furthermore, the recent research has shown that optimized charging on the basis of using AI algorithms can double battery cycle life in grid-scale storage applications, along with maintaining 95% of original capacity [4]. Resultantly, these improvements have the ability to make tremendous implications by reducing the lifetime costs on the economics of renewable energy storage.

5.3.4 AI for Energy Storage Efficiency and Lifespan Extension

5.3.4.1 Adaptive Control Strategies

Artificial intelligence has introduced new adaptive control techniques for energy storage systems, which have transitioned these systems' reaction to dynamic operating conditions. In contrast to conventional static control methods, AI-powered adaptive systems use complex machine learning architectures for continuously learning and tuning their parameters in real time.

Moreover, Deep reinforcement learning (DRL) architecture enables storage systems to automatically create the most suitable control algorithm by interacting with their surroundings and receiving feedback on the performance results, which has been proven effective. Additionally, these AI controllers can also analyze a wide range of input factors such as Prices of current energy, load estimates, expected renewable generation, and comprehensive data of battery state, which resultantly make decisions about power flow in milliseconds.

These adaptive strategies are so powerful due to their ability to handle complex, multidimensional optimization problems that were troublesome for the traditional approaches. For example, an AI controller concurrently optimizes for avoiding degradation impacts, lowering electricity costs, upholding grid frequency regulation commitments, and maintaining adequate reserve capacity while taking future energy supply and demand uncertainties into consideration. In contrast to traditional methods, neural network-based controllers have shown the capacity to learn nonlinear relationships between operating parameters and long-term degradation effects, resulting in control strategies that can increase battery lifespan by 40% or more.

5.3.4.2 Fault Detection and Predictive Maintenance

AI-based fault detection and predictive maintenance systems represent a dramatic advancement in energy storage reliability and safety, as these systems employ a combination of supervised, unsupervised, and semi-supervised machine learning techniques for monitoring the hundreds of operational parameters and detecting the anomalies that may result in developing faults. Deep neural networks process time series data from voltage, current, temperature, and impedance measurements for identifying patterns that precede common failure modes such as thermal runaway, lithium plating, or separator degradation. The most advanced systems incorporate graph neural networks for modeling relationships between different battery cells in a pack, enabling early detection of developing imbalances or weak cells that could lead to cascading failures.

Predictive maintenance algorithms estimate the remaining useful life (RUL) and predict specific maintenance needs before failures occur. These systems combine the physical models with data-driven approaches by using techniques of physics-informed neural networks, which create digital twins that simulate the aging processes under various operating conditions. Additionally, the analysis of historical degradation patterns across thousands of cycles enables AI models to recommend the schedules of optimal maintenance, operational adjustments, and predict capacity fade trajectories for mitigating some specific degradation processes. Therefore, the implementation of these systems has shown remarkable results in the field deployments by reducing the unplanned downtime up to 75% along with decreasing maintenance costs by 30–50% when compared to the traditional scheduled maintenance approaches.

In addition, the recent developments in this field also involve the use of explainable AI (XAI) techniques, which provide interpretable diagnostics and AI implementations for real-time monitoring without depending on the cloud and hybrid models that combine data from multiple sensing techniques such as Thermal imaging, Ultrasound, and Gas sensors. Moreover, these AI systems are increasingly adopted as a crucial part of battery health management platforms, which in turn continuously learn from wide data along with preserving the protected information of individual operators. As energy storage systems are becoming complex and vast with time, these AI-powered predictive maintenance is playing the role of not just an optimization tool but also a critical safety system that mitigates the catastrophic failures and ensures reliable operation throughout the life of the designed system. Eventually, the integration of these technologies is enhancing the standards for energy storage reliability, which ultimately improves the economic viability of storage deployments on a large scale.

5.4 AI in Thermal Energy Storage (TES) Systems

5.4.1 Overview of TES and AI Integration

Thermal energy storage systems play a vital role in enhancing the efficiency and sustainability of energy infrastructure by storing thermal energy, thus eradicating the mismatch between energy generation and demand. This system comprises various technologies, including sensible heat storage, latent heat storage, and thermochemical storage. Integrating these in HVAC systems, solar thermal power plants, and heating proved to be vital for energy demand management and peak load reduction. Similarly, during low-generation periods, efficient thermal energy storage ensures grid stability. However, these systems are complex and exhibit non-linear behaviors, this is where AI emerges as a useful application.

The integration of AI into thermal energy systems changed the perspective of how these systems are designed, controlled, and optimized. AI models can learn from the historical and real-time data provided to the model, thus using intelligent decision-making to improve the overall efficiency of the system. These techniques, particularly Machine Learning and deep learning, are being employed to predict energy loads, optimize storage strategies, and enhance fault detection and diagnosis. Hybrid AI models combining optimization algorithms like genetic algorithms (GAs) or particle swarm optimization (PSO) with ML methods are being used for multi-objective optimization of TESS operations. Thus, these AI-enabled solutions help in reducing energy waste, operational costs, and carbon emissions, hence meeting global sustainability goals [5].

5.4.2 AI for Thermal Load Forecasting and Energy Dispatch

Thermal load forecasting plays an important role in managing and optimizing Thermal energy storage operations. Using AI algorithms such as artificial neural networks, support vector machines, and deep learning architectures, forecasting thermal loads is possible, provided the historical data, weather parameters, and other important system variables. Accurate forecasting of the Thermal energy systems enables charging and discharging at optimal times, thus minimizing the energy wastage and operational costs too. These models can learn complex relations between input parameters and thermal demand, thus outperforming the previous statistical forecasting methods. Similarly, in buildings and district heating systems, load profiles are highly variable: using AI-driven forecasting systems in these buildings will enhance the efficiency and response rate of thermal energy storage systems.

Similarly, AI has a significant contribution in energy dispatch optimization by ensuring that thermal energy is used in the most efficient and cost-effective manner. Considering real-time data and environmental conditions of a specific area, AI-based dispatch systems decide when to store and release the energy and how much energy is to be stored and released. Using techniques such as fuzzy logic and a hybrid optimization approach, dispatch schedules can be created that balance load demands, thermal comfort, energy prices, and system constraints altogether. For instance, reinforcement learning can enable a TES controller to learn optimal dispatch strategies over time through trial-and-error interactions with its environment.

5.4.3 Dynamic Pricing and Demand Response Optimization

Role of AI is pivotal in dynamic pricing and demand response optimization as energy costs fluctuate based on supply-demand dynamics. AI systems, if integrated with thermal energy storage, can predict price variations and shift energy usage from peak hours to off-peak hours, store energy when it's cheaper, and dispatch energy during peak hours. This will lower operational costs and also contribute to the stability of the grid by flattening peak demands. Concludingly, the integration of AI in Thermal energy systems across these three dimensions i.e., forecasting, dynamic pricing response, and smart control, is a step toward intelligent and sustainable energy systems.

5.4.4 AI Applications in Low, Medium, and High-Temperature TES

AI applications in low, medium, and high temperature thermal energy systems vary depending upon temperature ranges, the type of application, and the technology. These temperature ranges can be broken down into three different levels: low temperature below 100°C, medium temperature between 100°C and 400°C, and high temperature above 400°C. An AI application for three different ranges is presented in Table 5.4.

TABLE 5.4

AI Applications and Usage for Different Temperature Ranges

Temperature Ranges	Typical Usage	AI Applications
Low	Residential/commercial buildings, HVAC, domestic hot water, space heating.	Thermal Load Forecasting, Smart HVAC Control, Energy Dispatch Optimization, Fault Detection & Maintenance, Integration with Smart Grids.
Medium	Industrial processes, solar thermal plants, district heating	Process Optimization, Predictive Maintenance, Energy Cost Minimization, Thermal Efficiency Monitoring, Renewable Integration
High	Concentrated Solar Power (CSP), metal processing, cement production, thermal battery systems.	Dynamic System Modeling, High-Precision Dispatch Control, Operational Safety Monitoring, Hybrid Energy System Optimization, Thermochemical Storage Management

5.4.5 Case Studies and Research Developments in Low, Medium, and High-Temperature TES

Across all temperature ranges, using AI, thermal energy systems are becoming more adaptive and economically efficient. A few recent research developments across all three temperature ranges are discussed in Table 5.5.

TABLE 5.5

Case Studies Regarding AI Integration

Temperature Ranges	Case Studies
Low	1. In Germany, ANN-based models were deployed to predict daily heating loads based on external temperature, occupancy patterns, and solar radiation. These models helped in optimizing the operation of water-based thermal storage tanks, resulting in up to 18% energy savings while maintaining indoor comfort. 2. In Sweden, reinforcement learning (RL) was implemented to manage air-to-water heat pumps with TES in homes under dynamic pricing.
Medium	1. In India, ANN models were used to manage thermal energy systems in food processing plants where steam and hot water are required. 2. In Spain, researchers at CIEMAT used support vector regression (SVR) to forecast thermal energy demands in a district heating network, improving the operational efficiency by over 15% comparatively.
High	Noor Energy 1 CSP project in the UAE employed model predictive control (MPC) enhanced with AI forecasting to manage molten salt charging/discharging. AI ensured thermal energy was stored during peak sunlight and dispatched optimally to sustain power generation during cloudy periods or after sunset—improving dispatch reliability by over 20%.

5.5 Challenges in AI-driven Energy Storage Systems

The integration of AI into energy storage systems, particularly thermal energy storage, plays an important role in enhancing the operational efficiency of the systems. However, there are a few challenges that need to be addressed. Different challenges while integrating AI with Thermal Energy Storage systems are discussed below.

5.5.1 Data Acquisition

In an AI system, model training depends upon consistent time-series data, contextual data integration such as weather, and multi-dimensional datasets to operate efficiently and reliably, while missing values or inconsistent data can halt the accuracy of the model and its training efficiency. In many thermal energy systems applications, due to these data inefficiencies, sensor coverage may be insufficient and difficult to maintain [6].

5.5.2 Model Transparency and Integration Complexity

In AI systems, particularly deep learning, decision-making process is not transparent and the explainability of the background processes is limited. In an energy infrastructure setting, this poses a barrier as human operators need to operate and trust the decision-making process. The lack of interpretability is of great concern where decisions affect large systems and critical operations, such as Power plants, Electrical grids, and Large thermal energy systems. Here, there can be serious consequences like financial loss and even safety hazards [7]. Similarly, as far as the integration complexity is concerned, AI requires continuous interactions with the existing control systems and sensors. As thermal energy systems are multi-physical i.e., (thermal, fluidic, and mechanical), their integration becomes complex, and this fusion of physics-based models is still being researched, and there is a research gap to be covered.

5.5.3 Generalization, Transferability and Data Privacy

AI systems are trained depending upon the time-series data provided, that do include different climate zone constraints. Thus, an AI model trained at a certain location for a certain system may not perform well when transferred to a different climate zone and for different thermal energy system configurations. Thus, the field of transfer learning in energy systems is still progressing and is still developing.

Thermal energy systems are exposed to cyber threats, including spoofing, data breaches, etc. This is because the thermal energy system is connected via IoT networks. The problem even worsens when this data is of a sensitive nature and adds privacy risks to it [8].

5.5.4 Robustness and Adaptability of AI Systems

The ability of AI models to maintain performance despite changes in the input data or sensor inaccuracies is known as the Robustness of the model. Thermal energy systems work in different environments where weather conditions are changing, and so are the user demands. Similarly, from time to time, equipment is degrading. Thus, a model that is trained on idealized conditions may not perform well during unexpected scenarios for which the data isn't provided. For example, a neural network that is trained to forecast building loads during a particular season, say winter, may not perform properly during a sudden heat wave. Similarly, AI models may become unstable during sensor faults or due to a delay in communication. So, a robust preprocessing mechanism is required [9].

Thermal Energy systems evolve due to the aging of components, different maintenance schedules, or different changes in external conditions such as utility pricing, so these systems are not static. Therefore, AI systems should be adaptive and evolving with respect to system behavior. Unfortunately, many AI implementations rely on models that have been trained offline, so they do not update over time and need to be trained explicitly. Now, the difficulty with training explicitly is that the operation becomes disruptive and is computationally expensive as well. To counter these, reinforcement learning is being employed along with meta learning approaches. Reinforcement learning agents can interact with Thermal Energy system environments and adapt to changing thermal loads and pricing conditions [10]. A few advantages of the robust and adaptable AI systems have been presented in Figure 5.2.

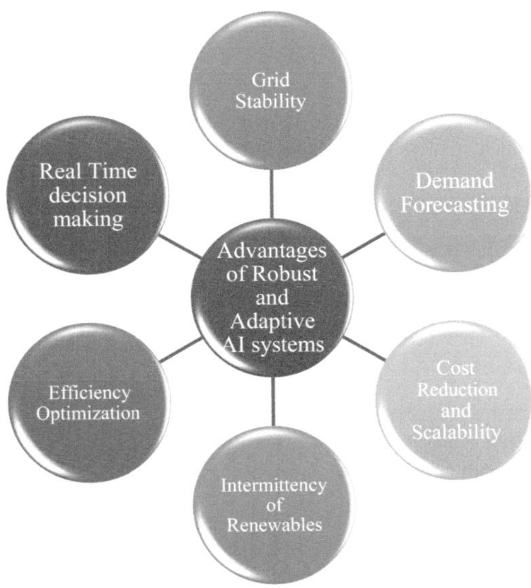

FIGURE 5.2
Advantages of robust and adaptive AI systems. (*Source*: Author.)

5.5.5 Hybrid Model as a Solution

Now, to enhance the robustness and adaptability of the AI systems, hybrid models combining physics-based modeling with data-driven AI techniques are being studied. The current research is focusing on integrating thermodynamic principles into AI architectures. This allows ensuring physical consistency while training a model [11].

5.6 Future Perspectives and Emerging Trends

Reinforcement learning and generative AI are the future advancements in AI that may unlock new potential in thermal energy storage systems. Under the uncertain energy market trends, reinforcement learning can self-optimize control strategies for Thermal energy storage systems and contribute toward grid flexibility and improve the operational efficiency [12].

Another emerging trend involves the integration of digital twins with AI models. By creating real-time virtual replicas of TES systems, operators can test optimization strategies, predict faults, and conduct scenario-based forecasting without disrupting physical operations. These AI-enabled digital twins could be particularly transformative in large-scale industrial TES settings, where predictive maintenance and system reliability are critical.

5.6.1 Integration with IoT and Blockchain for Smart Energy Systems

The integration of Artificial Intelligence (AI), the Internet of Things (IoT), and blockchain technology is marking new ways for the development of intelligent and autonomous Thermal Energy Storage (TES) systems. IoT devices—such as embedded sensors and actuators—enable continuous, real-time monitoring of key thermal parameters, including fluid temperature, heat flux, and energy transfer rates. This data-rich environment enhances the accuracy and responsiveness of AI models in forecasting energy demand and optimizing system performance [13].

Blockchain provides a secure, transparent, and decentralized platform for energy transactions. Within smart microgrids, blockchain facilitates peer-to-peer thermal energy trading that is supported by smart contracts that execute energy exchange agreements automatically based on AI-generated forecasts of supply and demand imbalances. Using this integrated approach strengthens system transparency, improves the operational efficiency of the system, and supports the creation of resilient, distributed energy ecosystems [14].

5.6.2 Policy and Regulatory Considerations for AI Adoption

While the integration of AI into TES systems holds significant promise, its large-scale deployment is currently constrained by regulatory and policy challenges. One of the primary barriers is the absence of standardized frameworks for AI governance, including data ownership, model transparency, and cybersecurity protocols. Without clear guidelines, stakeholders may face difficulties in ensuring trust, safety, and interoperability across AI-enabled energy systems [15].

To facilitate wider adoption, policymakers must develop comprehensive regulations that promote responsible AI deployment in the energy sector. These should include the establishment of data interoperability standards, requirements for explainable AI models, and mechanisms for auditing AI-driven decisions in critical infrastructure [16].

References

1. Zou, B., Wang, H., Zhang, T., Xiong, M., Xiong, C., Sun, Q., & Ruan, H. (2023). A deep learning approach for state-of-health estimation of lithium-ion batteries based on differential thermal voltammetry and attention mechanism. Frontiers in Energy Research, 11, 1178151.
2. Li, K., & Chen, X. (2025). Machine learning-based lithium battery state of health prediction research. Applied Sciences, 15(2), 516.
3. Park, S., Pozzi, A., Whitmeyer, M., Perez, H., Joe, W. T., Raimondo, D. M., & Moura, S. (2020, August). Reinforcement learning-based fast charging control strategy for li-ion batteries. In 2020 IEEE Conference on Control Technology and Applications (CCTA) (pp. 100–107). IEEE.
4. Ukoba, K., Olatunji, K. O., Adeoye, E., Jen, T. C., & Madyira, D. M. (2024). Optimizing renewable energy systems through artificial intelligence: Review and future prospects. Energy & Environment, 35(7), 3833–3879.
5. Entezari, A., Aslani, A., Zahedi, R., & Noorollahi, Y. (2023). Artificial intelligence and machine learning in energy systems: A bibliographic perspective. Energy Strategy Reviews, 45, 101017.
6. Mocanu, E., Nguyen, P. H., & Gibescu, M. (2018). Deep learning for power system data analysis. In Big Data Application in Power Systems (pp. 125–158). Elsevier.
7. Forootan, M. M., Larki, I., Zahedi, R., & Ahmadi, A. (2022). Machine learning and deep learning in energy systems: A review. Sustainability, 14(8), 4832.
8. Said, O., & Masud, M. (2013). Towards internet of things: Survey and future vision. International Journal of Computer Networks, 5(1), 1–17.
9. Faritha Banu, J., Atul Mahajan, R., Sakthi, U., Kumar Nassa, V., Lakshmi, D., & Nadana kumar, V. (2022). Artificial intelligence with attention based BiLSTM for energy storage system in hybrid renewable energy sources. Sustainable Energy Technologies and Assessments, 52, 102334.
10. Abdi, N., Albaseer, A., & Ab-dallah, M. B. T.-R. M. in ES and ES. (2024). The Integral Role of AI in Energy and Smart Grid Technologies. Elsevier.
11. Ali, A. N. F., Sulaima, M. F., Razak, I. A. W. A., Kadir, A. F. A., & Mokhlis, H., 2023. Artificial intelligence application in demand response: Advantages, issues, status, and challenges. IEEE Access, 11, 16907–16922. https://doi.org/10.1109/ACCESS.2023.3237737

12. Sun, Q., Fu, Y., & Lin, H. et al. (2022). A novel integrated stochastic programming-information gap decision theory (IGDT) approach for optimization of integrated energy systems (IESs) with multiple uncertainties. Applied Energy, 314, 119.
13. Bakar, A. A., Yussof, S., Ghapar, A. A. et al. (2024). A review of privacy concerns in energy efficient smart buildings: Risks, rights, and regulations. Energies, 17(5), 977.
14. Andoni, M., et al. (2019). Blockchain technology in the energy sector: A systematic review of challenges and opportunities. Renewable and Sustainable Energy Reviews, 100, 143–174.
15. European Commission. (2020). Ethics guidelines for trustworthy AI. https://digital-strategy.ec.europa.eu
16. IEA. (2021). AI in energy: Opportunity or threat? International Energy Agency. https://www.iea.org

6

Sustainable Synergy with AI and Energy Storage

Muhammad Ihsan-Ul-Haq and Saad Naseem

6.1 Introduction

Thermal energy storage (TES) is a key enabler for deep decarbonization because it buffers variability across renewable electricity, industrial heat, and building thermal loads. The central opportunity for artificial intelligence (AI) in TES is not merely automation, but accelerated decision-making across scales from materials selection, such as salts, phase-change materials, and redox couples, to device design, including heat transfer intensification, and reactor architectures [1]. In practical terms, AI methods, such as machine learning, data analytics, and intelligent control, are applied to improve the efficiency, performance, and lifetime of energy storage systems, which in turn enable greater adoption of renewable energy and more reliable, low-carbon power systems. The outcome is a virtuous cycle, i.e., AI makes energy systems smarter and more adaptive, facilitating higher renewable penetration and better resource utilization, while advanced energy storage provides the flexibility and resiliency that AI-driven control strategies require to maintain stability in sustainable energy grids. This synergy is increasingly seen as pivotal for the energy transition, as traditional grid management approaches struggle to accommodate the variability of renewables and the complexity of modern power demand [2]. The ability of AI-based models to process massive data streams and optimize complex systems in real-time offers a powerful tool to manage this complexity. For example, AI techniques could be used to forecast solar and wind power generation with high accuracy, optimize the materials for TES systems, and coordinate distributed energy resources across the grid.

Long-duration heat storage and efficient thermal management unlock intermittent renewable (solar/wind) penetration and safer, longer-lived TES systems. TES spans sensible, latent, and thermochemical energy storage pathways, each balancing energy density, temperature range, cost, and integration complexity [3]. There is growing interest in storing energy as heat using abundant materials like salt, bricks, and sorbents like silica gel and zeolites.

DOI: 10.1201/9781003516132-6

Multiple demonstrations across concentrated solar power (CSP), district heating, and industrial heat delivery confirm TES feasibility at scale. However, their broad deployment remains constrained by three coupled bottlenecks: (i) materials that simultaneously meet temperature window, stability, and cost targets; (ii) component designs that raise charge/discharge power without sacrificing durability; and (iii) control strategies that translate forecasting and operational constraints into safe, value-maximizing dispatch. These are precisely the domains where AI can significantly contribute when coupled to physics-informed models and validated data.

6.2 Fundamentals of TES

Sensible TES stores heat through temperature rise in a solid or liquid medium, making it structurally simple and scalable. The design trade-off is typically between (i) operating temperature range and freeze/solidification risk for salts, (ii) heat transfer rates and thermal gradients for solids, and (iii) long-term chemical compatibility with containment materials. For instance, nitrate salt mixtures, such as $NaNO_3/KNO_3$ and solar salts, offer potential for TES due to depression of melting temperatures upon their nanoconfinement in nanoporous anodic aluminum oxide [4], as shown in differential scanning calorimetry (DSC) plots in Figure 6.1. AI could add value primarily by enabling rapid property screening, which includes composition–melting behavior and corrosion propensity, and by learning reduced-order models that accelerate design exploration under multi-objective constraints, such as cost, heat rate, and storage density.

Latent TES uses phase-change materials (PCMs) to provide near-isothermal storage and can deliver heat at a controlled temperature during phase transition. The central design requirements include (i) transition temperature and enthalpy matched to the application; (ii) cyclic stability and suppression of supercooling/phase segregation; and (iii) heat transfer intensification due to the intrinsically low thermal conductivity in many PCMs. Therefore, AI is most effective when trained on curated PCM property datasets, such as the slPCMlib database [5], and used to solve inverse design tasks by proposing compositions that meet target melting point, enthalpy, and stability constraints simultaneously. Figure 6.2 [6] can be employed as a practical design map in which AI models can be trained to recommend candidate PCM families once a target temperature window and minimum volumetric enthalpy are specified, then refined by constraints such as compatibility, subcooling behavior, and cyclic durability.

Thermochemical energy storage (TCES) stores heat through reversible reactions, offering high energy density and low standby losses. Their requirements include a high energy content per mass (often two- to four-fold sensible heat) and the ability to store for long durations without insulation.

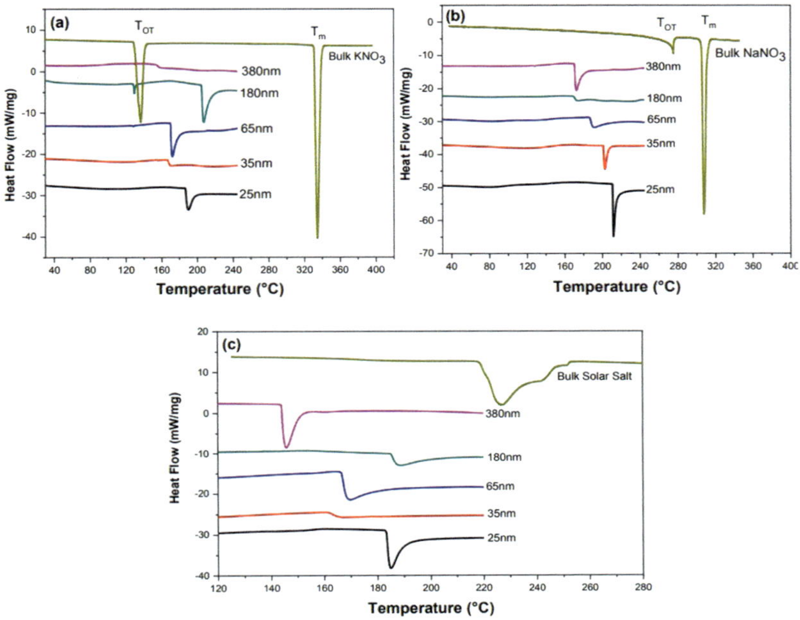

FIGURE 6.1
DSC measurements of (a) KNO_3, (b) $NaNO_3$, and (c) solar salts infiltrated with 25, 35, 65, 180, and 380 nm Pore diameters of anodic aluminum oxide. (Reproduced with permission from [4].)

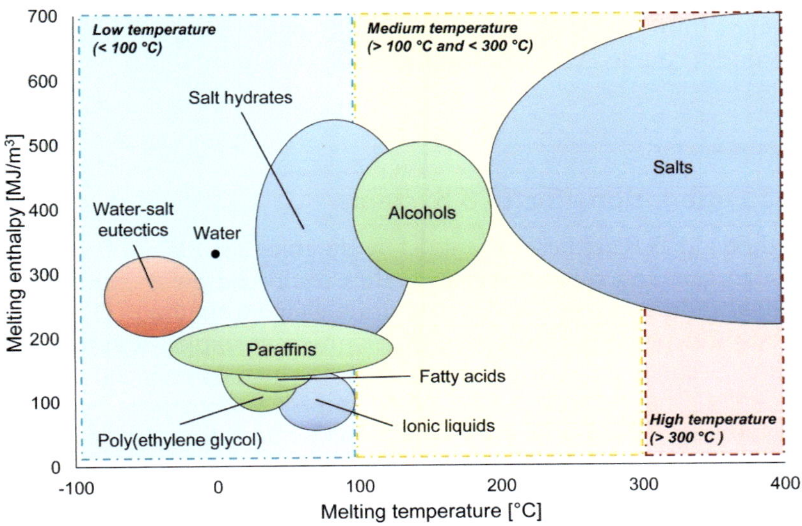

FIGURE 6.2
Schematic of PCM melting temperature and volume-specific enthalpy ranges. (Reproduced with permissions from [6].)

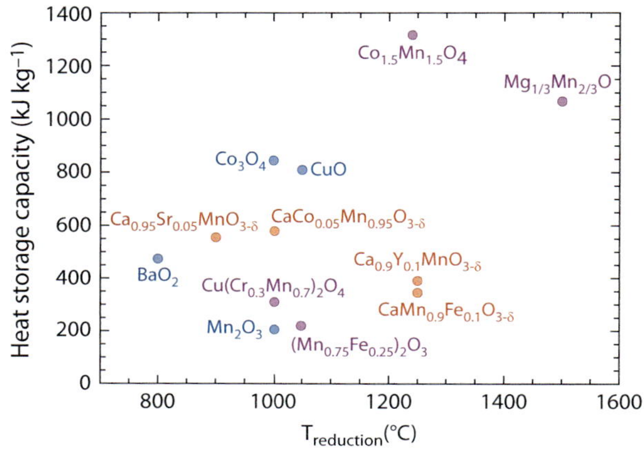

FIGURE 6.3
Heat-storage capacity versus reduction temperature for some promising TCS redox oxides. (Reproduced with permission from [7].)

The dominant constraints are kinetic limitations, mass and heat transport within reactive beds, and long-term reversibility under multiple cycles due to sintering, phase changes, and volatile loss. For instance, metal oxide redox pairs [7] can be used for TCES, as shown in Figure 6.3. From an AI perspective, the most impactful opportunities include (i) screening reaction families for feasible operating windows and energy density; (ii) inferring kinetic parameters from sparse experiments; and (iii) optimizing reactor/control co-design under coupled transport and reaction constraints.

6.3 AI Foundations for TES Systems

AI methods in TES are best organized by the role they play in the workflow. First, surrogate and reduced-order models could compress expensive computational fluid dynamics (CFD) and finite element method (FEM) simulations into fast predictors that enable design-space exploration and inverse problems. Second, physics-informed learning, such as artificial neural network (ANN), and hybrid models could improve robustness by enforcing conservation laws and boundary conditions while fitting data, as shown in Figure 6.4 [8]. Graph-based learning is useful when TES systems are naturally represented as networks, such as components, flow paths, an spatial discretization, while sequence models support anomaly detection and forecasting in operational data streams. Reinforcement learning can improve dispatch policies when paired with safety constraints or model-predictive control to avoid unsafe exploration, making it better suited for simulated training and constrained deployment rather than unrestricted real-time learning. Third,

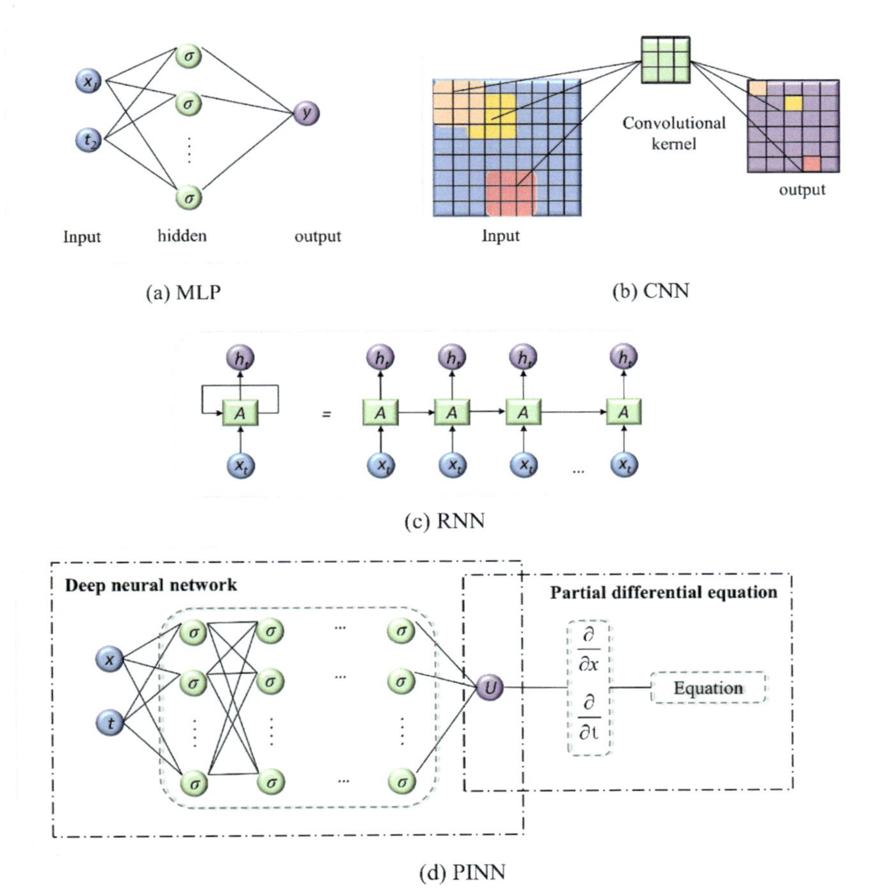

FIGURE 6.4
Schematic diagram of a typical ANN structure. (Reproduced with permission from [8].)

forecasting and control algorithms could translate uncertain loads, prices, and renewable availability into safe charge–discharge decisions under operational constraints.

6.4 AI for Materials Discovery and Property Prediction in TES Systems

AI is expediting the search for PCMs with tailored melting points and high latent heats. Data-driven models, including regression and classification, predict key PCM properties from molecular and/or formulation features.

AI contributes to TES materials discovery through three typical reusable workflows. First is property prediction, in which supervised learning maps molecular or compositional descriptors to melting point, enthalpy, heat capacity, and stability indicators, enabling rapid down-selection. Second is physics-accelerated modeling in which reduced-order and surrogate models predict phase-change dynamics or molten-salt thermophysics at a fraction of CFD or molecular dynamics (MD) cost, supporting inverse design. The third workflow is optimization in which Bayesian or evolutionary strategies navigate composition and geometry spaces under explicit constraints, such as temperature window, stability, and manufacturability, prioritizing candidates for experimental validation.

For instance, AI could assist in designing PCM composites that overcome low thermal conductivity. Surrogate models, trained on computational or experimental results, allow rapid screening of combinations, such as PCM with graphene or aluminum foam at various fractions, for effective conductivity and specific heat. A rapid and accurate reduced-order model was developed (see Figure 6.5) combining proper orthogonal decomposition (POD) and Bayesian neural networks (BNN) to efficiently predict transient melting behavior in shell-and-tube latent TES devices, achieving significant computational speedup and reliable accuracy across varying operating conditions compared to traditional high-cost CFD simulations [9].

For high-temperature sensible heat storage, multi-component molten salts including nitrates, chlorides, and carbonates are being optimized for wider thermal stability and lower freezing points. AI could aid this by building predictive models of salt properties, such as melting point, decomposition temperature, and specific heat, based on the salt composition. An overview of a typical workflow of ML-based studies on molten salt is illustrated in Figure 6.6a [10]. The microstructure and thermal properties of $NaCl–CaCl_2$ eutectic salt were investigated using two data collection methods and three machine learning potentials, evaluating partial radial distribution functions, coordination numbers, and key thermal properties, while comparing deep potential (DP) and neuroevolution potentials (NEP) to identify efficient and accurate machine learning approaches for modeling multicomponent molten salt phase change materials, as shown in Figure 6.6b [11]. Besides, a machine learning interatomic potential capable of accurately predicting thermophysical properties of 11-cation chloride molten salts with near-DFT accuracy, named as SUPERSALT, offered excellent transferability across compositions and enabled accelerated discovery of optimal salt mixtures through Bayesian optimization, thus providing a powerful platform for simulating and optimizing complex molten salt systems for clean energy applications [12]. The complete workflow for their training potential is shown in Figure 6.7.

Nanoparticle-enhanced molten salts are an emerging area in which adding a small amount of ceramic or carbon nanoparticles can substantially increase the specific heat capacity or thermal conductivity of salts. However, ensuring their stability by avoiding agglomeration and corrosion is challenging,

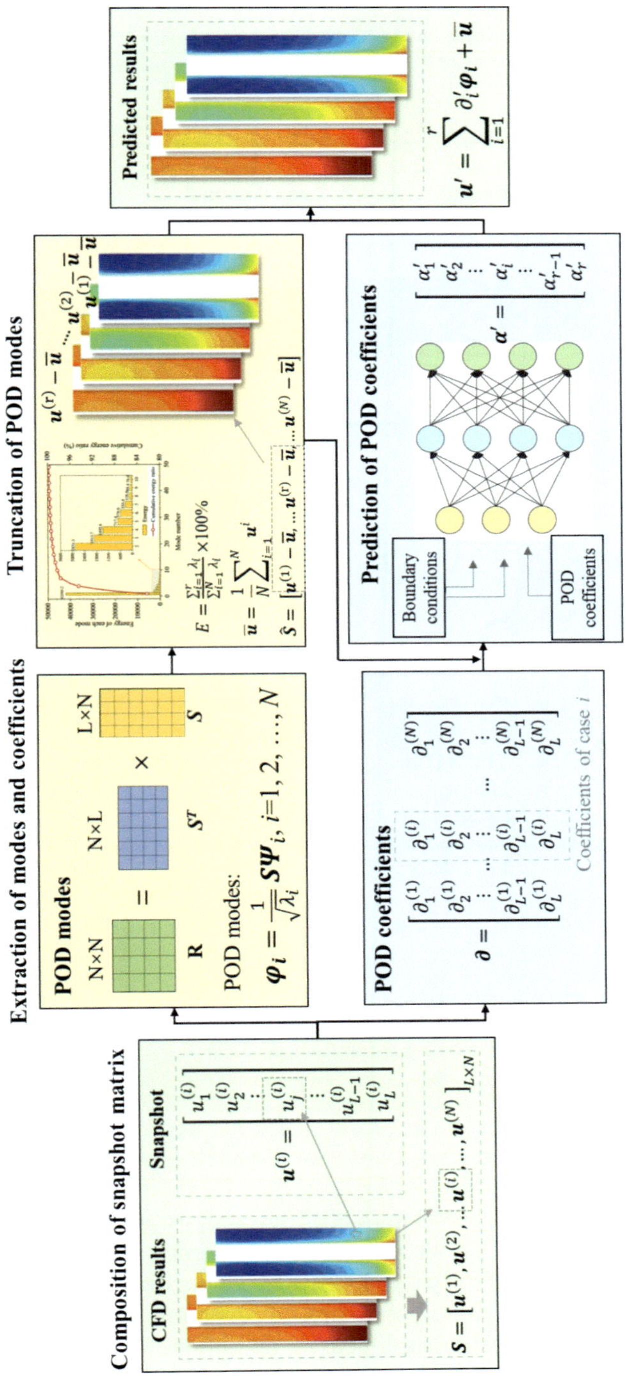

FIGURE 6.5
Flowchart of flow field reconstruction based on POD and bayesian neural network. (Reproduced with permission from [9].)

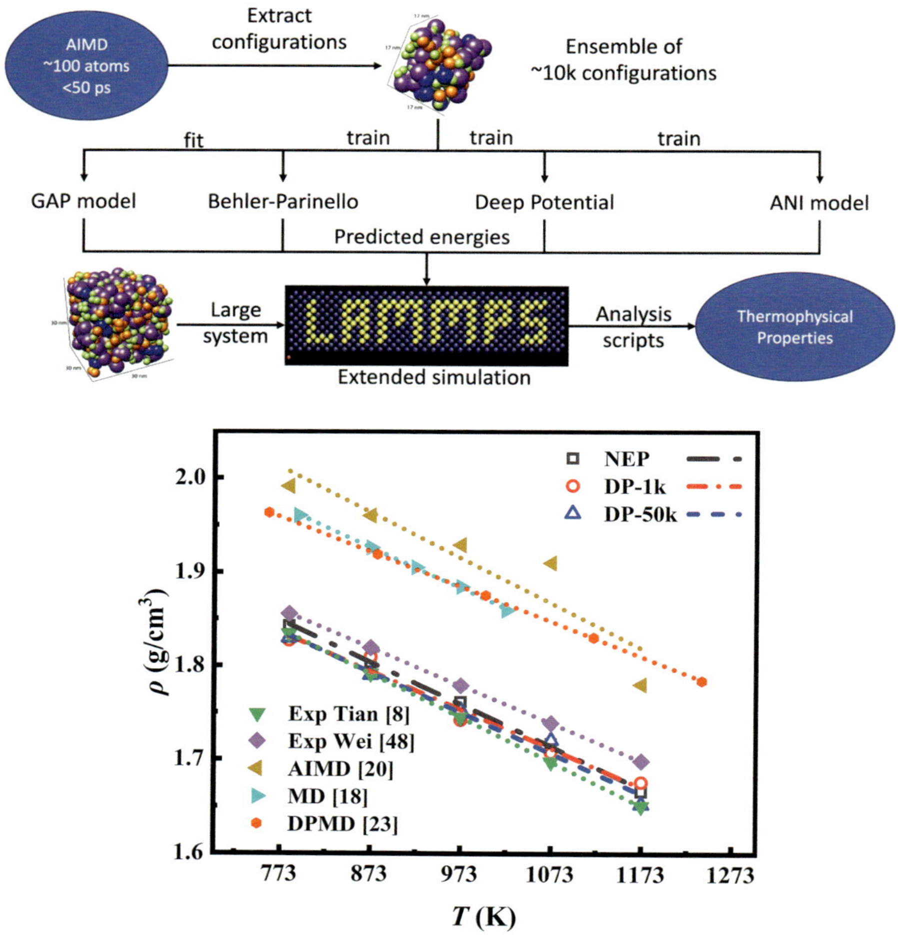

FIGURE 6.6
(Top) overview of typical methods used in machine learning based studies of molten salts. Reproduced with permission from [10]; (Bottom) densities of NaCl–$CaCl_2$ eutectic salt obtained by different methods. (Reproduced with permission from [11].)

where AI classification models could be used to predict which nanomaterials and surface treatments are likely to remain dispersed and effective, respectively.

The search for reversible reactions with high energy density and cyclability is complex in TCES. AI is being leveraged to screen material candidates and model their kinetics. For example, five ML algorithms were applied to model and predict the dynamic behavior of a reversible TCES system based on NH_4NO_3 and KOH, identifying k-nearest neighbors (KNN) and Bayesian Ridge as the most accurate methods for regression modeling, while extreme

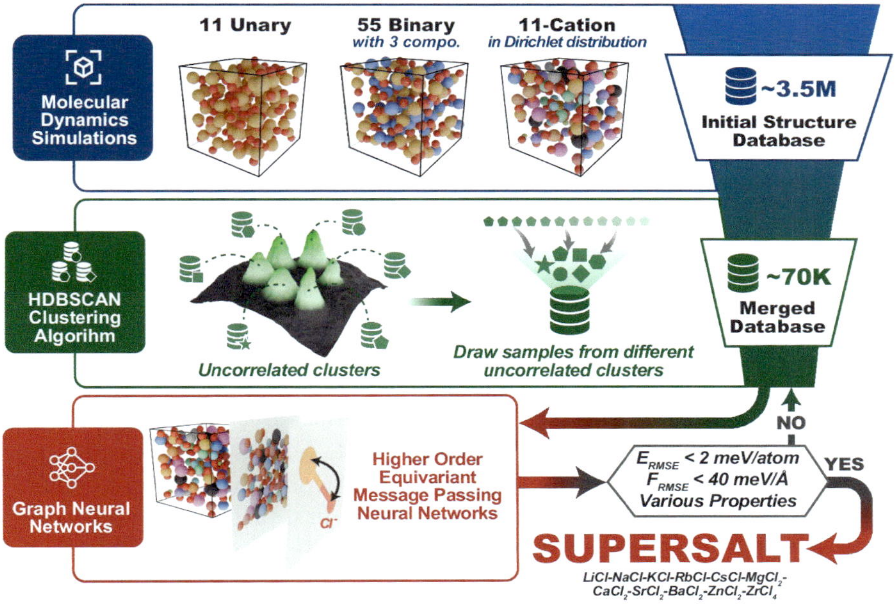

FIGURE 6.7
A comprehensive workflow for training SuperSalt potential. (Reproduced with permission from [12].)

gradient boosting (XGBoost) effectively forecasted future temperature trends from time-series data, significantly reducing computational costs compared to traditional physics-based approaches [13]. Besides, ML-based heat load forecasting was integrated by using support vector regression, regression trees, and long short-term memory networks with model predictive control (MPC) to optimize seasonal TCES in the district heating network, demonstrating that MPC could significantly improve heat demand management, reduce waste heat, and outperform rule-based control, particularly under moderate supply capacities [14].

Open datasets and interoperable data standards can accelerate AI-based TES materials discovery, but the limiting factor remains curated, TES-specific benchmark datasets that include both materials properties and cyclic performance outcomes. Future value could come less from listing repositories and more from harmonizing measurement protocols and uncertainty reporting so that models trained across labs remain comparable and transferable. Researchers are increasingly releasing code as well. For example, a GitHub repository with the trained ML models and an interactive PCM selection tool is reported [15]. Such open science practices, coupled with the breadth of available data, could enable reproducibility and faster progress in the TES field.

6.5 Challenges and Outlook

Despite the progress, significant challenges remain at the intersection of AI and TES as outlined below:

i. AI models for TES materials, device optimization, and thermal management often rely on small or proprietary datasets. This hinders reproducibility and cross-comparison between methods. Unlike fields such as image recognition, TES lacks open benchmark datasets for PCM properties or molten salt stability. Establishing curated, standardized datasets that include thermophysical properties, cycling performance, and system-level operation will be critical for robust training, validation, and fair benchmarking of AI approaches.

ii. Most existing AI applications still treat thermal processes as black boxes. While physics-informed neural networks and reduced-order models have emerged, deeper coupling of first-principles with data-driven learning is needed to capture multi-phase flow, radiative heat transfer, and complex thermochemical reactions. Future work should advance hybrid digital twins that dynamically update their physics modules using AI, enabling accurate prediction across operating ranges while reducing computational cost.

iii. In safety-critical domains such as industrial thermal reactors, AI must operate with transparency, reliability, and override capabilities. Current black-box models risk misinterpretation and automation bias. Embedding uncertainty quantification, interpretable AI layers, and rule-based safety constraints will be essential. Therefore, human-in-the-loop frameworks, where operators supervise and refine AI decisions, could build trust and accelerate adoption.

iv. AI-designed TES materials and systems often face challenges in manufacturability and real-world durability. For instance, PCM composites optimized in silico may phase-separate after repeated cycling, and intricate fin geometries suggested by generative models may be impractical to fabricate. Bridging simulation and industrial practice requires AI frameworks that incorporate manufacturability, cost, and lifecycle assessment constraints directly into the optimization loop. Techno-economic analyses informed by AI can also highlight trade-offs between capital expenditure, operational efficiency, and emissions reduction.

v. Finally, TES integration depends on supportive policies and market mechanisms. AI can help quantify the value of TES flexibility in electricity and heat markets, but widespread deployment will require standardized performance metrics and market signals, such as carbon pricing and storage incentives. Coordinated efforts between technologists, policymakers, and AI researchers would be necessary to accelerate TES adoption in grids, industries, and buildings.

In conclusion, AI can accelerate discovery of materials for TES by shortening materials-to-system design loops, improving predictive fidelity through physics-consistent learning, and enabling safe, forecast-informed control. Near-term priorities include (i) benchmarking datasets for TES materials and cycling durability; (ii) validating hybrid digital twins for phase-change and thermochemical systems; and (iii) deploying frameworks that pair optimization with uncertainty and safety constraints. Progress along these axes will determine how rapidly TES can scale as a reliable decarbonization asset.

Acknowledgements

This publication is based upon work supported by King Fahd University of Petroleum & Minerals. M. I.-U.-H. Acknowledge the Interdisciplinary Research Center for Sustainable Energy Systems for the support received under Grant No. INSE 2517. S.N. acknowledges the support from the Joint BioEnergy Institute, U.S. Department of Energy, Office of Science, Biological and Environmental Research Program under Award Number DE-AC0205CH11231 with Lawrence Berkeley National Laboratory. The United States Government retains and the publisher, by accepting the article for publication, acknowledges that the United States Government retains a nonexclusive, paid-up, irrevocable, worldwide license to publish or reproduce the published form of this manuscript, or allow others to do so, for United States Government purposes. Any subjective views or opinions that might be expressed in the paper do not necessarily represent the views of the U.S. Department of Energy or the United States Government.

References

1. T. Lombardo, M. Duquesnoy, H. El-Bouysidy, F. Årén, A. Gallo-Bueno, P.B. Jørgensen, A. Bhowmik, A. Demortière, E. Ayerbe, F. Alcaide, M. Reynaud, J. Carrasco, A. Grimaud, C. Zhang, T. Vegge, P. Johansson, A.A. Franco, Artificial intelligence applied to battery research: Hype or reality? Chem. Rev. 122 (2022) 10899–10969. https://doi.org/10.1021/acs.chemrev.1c00108
2. M. Mansoor, F. Grimaccia, S. Leva, M. Mussetta, Comparison of echo state network and feed-forward neural networks in electrical load forecasting for demand response programs, Math. Comput. Simul. 184 (2021) 282–293. https://doi.org/10.1016/j.matcom.2020.07.011

3. R. Mabrouk, H. Naji, A.C. Benim, H. Dhahri, A state of the art review on sensible and latent heat thermal energy storage processes in porous media: Mesoscopic simulation, Appl. Sci. 12 (2022) 6995. https://doi.org/10.3390/app12146995
4. M.G. Yazlak, Q.A. Khan, M. Steinhart, H. Duran, Melting temperature depression and phase transitions of nitrate-based molten salts in nanoconfinement, ACS Omega. 7 (2022) 24669–24678. https://doi.org/10.1021/acsomega.2c02536
5. T. Barz, A. Bres, J. Emhofer, slPCMlib: A modelica library for the prediction of effective thermal material properties of solid/liquid phase change materials (PCM), Model. Conf. (2022) 63–74. https://doi.org/10.3384/ecp19363
6. A.I. Garivalis, D. Rossi, M. Seggiani, D. Testi, Beyond water: Physical and heat transfer properties of phase change slurries for thermal energy storage, Cell Rep. Phys. Sci. 5 (2024) 101905. https://doi.org/10.1016/j.xcrp.2024.101905
7. A.J. Carrillo, J.M. Serra, Recent progress on redox materials for high-temperature thermochemical heat storage, adv, Energy Sustain. Res. 6 (2025) 2400317. https://doi.org/10.1002/aesr.202400317
8. R. Mao, Y. Lan, L. Liang, T. Yu, M. Mu, W. Leng, Z. Long, Rapid CFD prediction based on machine learning surrogate model in built environment: A review, Fluids. 10 (2025) 193. https://doi.org/10.3390/fluids10080193
9. S. Shen, C. Wu, F. Duan, Machine learning for predicting the PCM melting process in a rectangular enclosure energy storage, AI Therm. Fluids. 1 (2025) 100001. https://doi.org/10.1016/j.aitf.2024.100001
10. T. Porter, M.M. Vaka, P. Steenblik, D. Della Corte, Computational methods to simulate molten salt thermophysical properties, Commun. Chem. 5 (2022) 69. https://doi.org/10.1038/s42004-022-00684-6
11. H. Tian, W. Dong, W. Zhang, C. Guo, Machine learning techniques to probe the properties of molten salt phase change materials for thermal energy storage, Cell Rep. Phys. Sci. 5 (2024). https://doi.org/10.1016/j.xcrp.2024.102042
12. C. Shen, S. Attarian, Y. Zhang, H. Zhang, M. Asta, I. Szlufarska, D. Morgan, SuperSalt: Equivariant neural network force fields for multicomponent molten salts system, Nat. Commun. 16 (2025) 7280. https://doi.org/10.1038/s41467-025-62450-1
13. S. Tasneem, H.S. Sultan, A.A. Ageeli, H. Togun, W.M. Alamier, N. Hasan, M.R. Safaei, Machine learning modeling of reversible thermochemical reactions applicable in energy storage systems, J. Taiwan Inst. Chem. Eng. 148 (2023) 104926. https://doi.org/10.1016/j.jtice.2023.104926
14. Z. Wei, P.W. Tien, J. Calautit, J. Darkwa, M. Worall, R. Boukhanouf, Investigation of a model predictive control (MPC) strategy for seasonal thermochemical energy storage systems in district heating networks, Appl. Energy. 376 (2024) 124164. https://doi.org/10.1016/j.apenergy.2024.124164
15. J. Woods, A. Mahvi, A. Goyal, E. Kozubal, A. Odukomaiya, R. Jackson, Rate capability and Ragone plots for phase change thermal energy storage, Nat. Energy. 6 (2021) 295–302. https://doi.org/10.1038/s41560-021-00778-w

7

Digital Twin Solutions for Heat Storage Applications

Ahmad Jamil

7.1 Introduction: Defining the Digital Twin

Global trends in the energy sector have three key driving forces: decarbonization, decentralization, and digitalization [1]. Digitalization can be concisely defined as the creation of a digital entity or tool and displaying it on a digital platform [2]. The digital transformation of the energy sector has been an increasing interest of industry and researchers for the last decade due to the progress in digital technologies and implementation of data-driven tools across various technological sectors [1]. One product of this increasing digitalization is the digital twin. Despite its growing popularity, the term digital twin is often misunderstood or loosely applied, and one of the main reasons is the term's treatment as a buzzword in digitalization narratives rather than a rigorously defined technology. Therefore, it is particularly important to have a conceptual definition of what a digital twin is, its purpose, and what separates it from conventional models before it can be created for any kind of system. The purpose of this chapter is to serve as a guideline from which readers can extract a more concrete definition of the digital twin concept and then apply it to heat storage technologies to create an effective digital twin. Instead of focusing on specific modelling techniques, emphasis is put on fundamentals and the correct conceptualization and contextualization of the digital twin. If the foundations are laid correctly, the specific modelling techniques and tools used to develop the digital twin for any kind of system are secondary in nature and are use-case specific. More often than not, digital tools like the digital twin end up not serving their desired purpose or utilization goals due to these conceptual mistakes.

The term digital twin was first coined by NASA in 2012 as: "An integrated multi-physics, multi-scale, probabilistic simulation of a vehicle or system that uses the best available physical models, sensor updates, fleet history, etc., to mirror the life of its flying twin [3]. Since then, the definition has evolved as the use of the concept spread to other technological domains, and now a thorough survey of the literature can yield several definitions of the concept.

DOI: 10.1201/9781003516132-7

TABLE 7.1

Various Definitions of the Digital Twins Found Across Literature

A digital twin refers to a virtual simulation that integrates multiple physical aspects, dimensions, and probabilities of a real-world system, using data gathered from different sources to mirror the behavior of its physical counterpart [4, 5].
A digital twin is a digital representation that serves as an equivalent to a tangible physical object [6].
A digital twin is a dynamic, real-time digital model of a system, process, or service, which allows for a complete evaluation and accurate representation of the actual system [7].
A digital twin is a collection of adaptive models built to digitally simulate the functioning of a physical environment. These models collect real-time data, forecast future scenarios, and suggest preventive measures to avoid system failures [8].
The primary aim of digital twin technology is to create a highly detailed and accurate digital counterpart of a physical entity [9].

Table 7.1 lists a few definitions from literature that define the concept of the digital twin in various ways. Combining all these definitions, the correct interpretation of the term is: the digital twin is a collection of models, sensors, and digital tools that link a physical entity – the "physical twin" and its environment – the "physical space", to a virtual representation of both, thus creating the "digital twin" in a "digital space". It is important to note here that this connection of the physical twin with the digital twin is ideally through its entire lifecycle. Thus, an important distinction is created from conventional models – the digital twin and space need to be able to evolve with time, mimicking the physical twin's evolution through the physical space and maintaining a dynamic connection. The digital twin is also distinct from tools like the Internet of Things and computer-aided design, where the former is purely concerned with the physical space and the latter is a pure digital representation. In fact, Internet of Things and computer-aided design can be seen as a subset of the digital twin.

In order to visualize this concept Figure 7.1 elaborates the digital twin as a layered concept, starting with the physical twin in the physical space, where

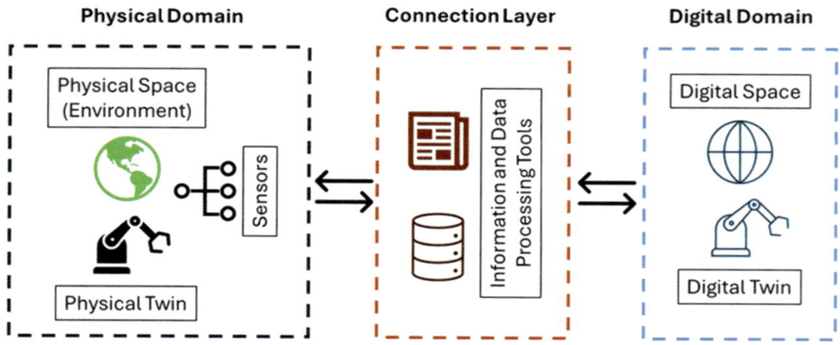

FIGURE 7.1
The three layers of the digital twin concept showing the major features of each layer and two-way communication between the layers. (*Source*: Author.)

adequate information is captured through sensors and then transferred by a real-time link into a digital space and twin, that not only mimic the physical domain in real-time, but evolve through the implementation of various modeling techniques. The connection layer between the physical and digital domains has a two-way flow and includes tools related to information and data management.

Once the concept is clear, the next step is to have a goal-oriented guideline in order to create an effective digital twin.

7.2 Creating an Effective Digital Twin

An ideal or complete digital twin not only has a real-time evolutionary link with the physical twin (termed as the twinning process), but serves the following goals through the entire product lifecycle [8]:

- **Design Stage:** The digital twin must capture the physical twin through the design process. It aids in conducting parametric studies to explore the design space, test for system irregularities, and simulate design iterations through potential future lifecycle scenarios to analyze degradations and predict performance.
- **Operations Stage:** The digital twin must simulate the physical twin's performance during operations by a continuous real-time link of information. It should have the ability to evolve with the physical twin as well. The digital twin's purpose at this stage can be divided into three subcategories:
    - i. Parametric studies for performance prediction, scenario analysis, degradation analysis, and evaluating the effects of upgrades or modifications to the physical system. The evolutionary capabilities of the twin serve a strong purpose here.
    - ii. Condition monitoring and fault diagnostics using the real-time link to ensure good maintenance and stable and effective operations of the physical twin.
    - iii. System control applications.

Both the design stage and operations stage macro-objectives can be broken down into further sub-objectives based on the specific technology the digital twin is being developed for. It is quite clear that a complete digital twin is not feasible to construct most of the time due to resource constraints. Therefore, an effective digital twin doesn't necessarily need to be able to fulfill all the above objectives. What is necessary for something to be called a digital twin is the twinning process. The real-time link and evolution are what complete the twining process. The term "real-time" is also subject to context. The time interval between packets of information between the real and digital is subjective and dependent on the use case. This increase in subjectivity or use-case dependency of terms requires that a step-by-step guide be developed to create a digital twin. The following steps present one such guideline that can be applied to any technology for the development of a digital twin.

1. Identify the system and define the environment for which the digital twin is to be developed. Similar to systems engineering approaches, a system boundary is defined, and so is an environment. By definition of environment, it is implied here that a selection of external factors that influence the system is made. Important questions that describe this step are: *What is the system, and what level of component fidelity is to be considered? Where is it located, and what environmental factors influence it? What resources does the system consume? What effect does it have on its environment?* These are necessary so that in the later stage, it is possible to identify where sensors are needed to capture relevant information from the physical domain.
2. Formulate a use-case for the digital twin. Questions like: *What life-cycle stage is the digital twin going to be used for? What use cases or applications during a specific lifestyle stage are to be analyzed by the digital twin?* This step helps in simplifying the complexity associated with a complete digital twin into an effective version.
3. Selection of sensors to capture adequate information from different system components and the environment. The sensory information should be adequate. This means that it captures information from the physical twin at the desired component level decided in the first step. It also captures the environment and ensures that all this information is sufficient for the use case decided in the second step. During this step, it is important to consider how the physical and digital domains will be linked, any possible noise influences, and define what is meant by the term real-time, i.e., at what time intervals packets of information reach the digital domain. This step also includes the selection of the required information and data processing tools needed to manage the two-way real-time link.
4. Once the above has been decided, the digital twin can be developed as a combination of models using various modeling techniques that serve the use-case. Irrespective of the type of modelling techniques used, the important part to include is the ability of the models to capture the evolution of the physical twin through its lifecycle. The models must be able to simulate "change" in the form of degradations, modifications, upgrades etc.

The above guidelines are summarized in Figure 7.2.

It is important to note here that the guidelines are not a silo-type process, but instead multiple steps can be done simultaneously, and each step can

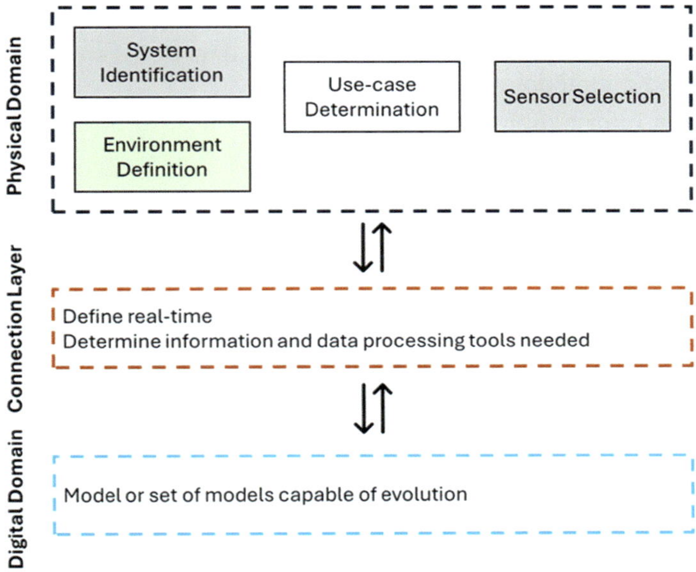

FIGURE 7.2
Summary and flowchart of the guidelines to develop an effective digital twin. (*Source*: Author.)

influence the other as needed. The following section applies the above guidelines to a case study of developing a digital twin for a heat storage system.

7.3 Effective Digital Twin for a Packed-Bed Thermal Energy Storage – A Case Study

The case study used to elaborate on the use of digital twins for heat storage applications is a packed-bed thermal energy storage system located at TU Wien [10]. The researchers have used a generic five-dimensional digital twin architecture for the system. In this section, the case study will be explained using the concept developed in Section 11.1 and the guidelines developed in Section 11.2. It will be shown that the fundamentals elaborated in these sections can be modified as needed to develop effective digital twins.

The three fundamental layers – the physical domain, connection layer, and digital domain have been broken down into a five-dimensional architecture in this case study as follows:

1. Physical domain or asset layer
2. The connection layer has been broken down into three sub-layers.

   a. An integration layer that classifies incoming data.
   b. A communication layer that transfers data between the physical and digital domains.
   c. An information layer that holds data and information management tools.
3. The digital domain is divided into two sub-layers.
   a. A functional layer that holds various models.
   b. A business layer that holds the business logic, analogous to the second step in the guidelines of Section 11.2.

The three layers and their sublayers are now elaborated with the guidelines as follows:

7.3.1 System and Environment Definition

The first step is to define the system and its environment. The packed-bed thermal energy storage system setup at TU Wein consists of an insulated vessel filled with gravel, a 15 kW_{th} electric heater, and a ventilation unit that is a blower with a maximum pressure height of approximately 0.4 bar. During charging, hot air is ventilated through the gravel, which increases its temperature and stores sensible thermal energy. For discharging, cold air is ventilated through the hot tank, and the heated air leaves the storage. The air temperature is controlled between ambient and 330°C. The environmental parameters of importance are ambient temperature and pressure. The temperature distribution across the length of the storage column is also of interest when it comes to deciding the fidelity of the system to be modelled.

7.3.2 Use-case Definition

The purpose of the digital twin here is to offer condition monitoring, diagnostic, control, and predictive services, i.e., services through the operations stage of the life cycle. For this purpose, the real-time thermodynamic performance of the system is required to be simulated by the digital twin.

7.3.3 Sensor Selection

16 thermocouples have been placed along the length of the storage system, with one at each end. In this way, the entire temperature distribution of the storage system and the inlet and outlet temperatures are captured. The inlet and outlet pressures are also measured along with the inlet mass flow. This arrangement of sensors captures the system and environment adequately within the desired fidelity and use case. From the desired use-case, real-time in this context means a very high sampling rate of data, i.e., very short intervals of time between information packets.

7.3.4 Model Selection

For the purposes of condition monitoring, diagnostics, control, and predictive services during operations, a fast model is required. Data-driven models are quite effective and popular for these applications due to their fast responses. They are system or machine-specific and are highly dependent on data quality. They are often called black box models as they don't require knowledge about a system's physics and working principles. Their effectiveness depends on the scope of available data, thus limiting their robustness. In contrast, physics-based models are equation-based models that incorporate knowledge about systems' workings from a physics perspective. They are often called white box models and can be quite robust. However, this comes with a high computational cost. The closer a physics-based model is to reality, the higher the computational cost. This limitation affects their practicality for certain fast response applications, and a trade-off must be made between model fidelity and computational cost.

For the given system, both a physics-based, and a data-driven model were developed, but the selected model was a grey box model. A grey box model is a combination of data-driven and physics-based approaches, leveraging the advantages of both techniques to achieve desired fidelity with efficient use of computational resources. The selected grey box model is primarily a data-driven model with physics-based "support" that enables predictions a few time steps ahead. This aids in predictive services, while the primary data-driven portion allows for condition monitoring.

7.3.5 Layer Integration and Development of the Digital Twin

Following the guidelines, the subjective elements have been defined in a goal-oriented way. The final step is to integrate everything in the layered architecture of the digital twin while ensuring a real-time link and evolvability. The five-dimensional layer system is developed under the guidelines as follows.

Figure 7.3 presents a complete summary of this process.

1. The 18 temperature sensors, 2 pressure sensors, and the mass flow sensor, along with the physical system, form the physical domain or asset layer.
2. Since the grey box model is primarily data-driven, with physics-based support, the classification, processing, storage, and management of data all fall into the three connection sublayers. The model's training data repository is also a part of this layer. This data repository can be updated, and thus the model can be trained at regular intervals to ensure its outputs evolve with the physical twin.
3. The data-driven grey box model, the business logic from the developed goals, and the use case all comprise the three sublayers of the digital domain.

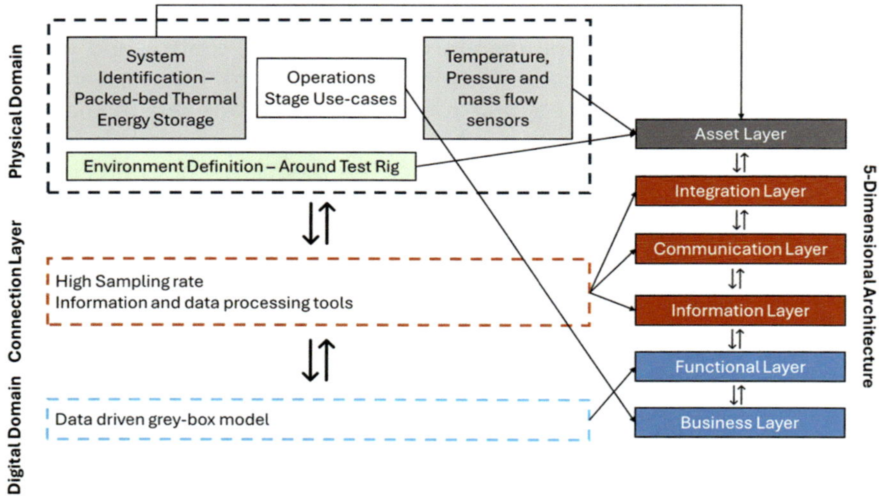

FIGURE 7.3
Summary and flowchart of the guidelines to develop an effective digital twin. (*Source*: Author.)

The end goal of a digital twin is to aid in decision-making throughout a system's lifecycle for increasing its efficacy. In the case study presented, the entire digital twin architecture is built around the packed-bed thermal energy storage system's operations, thus aiding operators to make decisions that improve the delivery of services. Digital twins do not replace conventional models and digital decision-making processes; instead, they systematically combine them to create a superior tool. This tool promises to improve efficacy by equipping all stakeholders with the insights needed for better decision-making regarding the system's value delivery.

References

1. Owens, Brandon. "The digital energy transformation." Boston: GE Power (2018).
2. Devi, Satyabati, and T. A. V. Murthy. "The need for digitization. Digital information resources & networks on India: Essays in honor of Professor Jagindar Singh Ramdev on his 75th birthday." New Delhi: UBS Publisherâ€™ s Distributors Pvt. Ltd (2005).
3. Glaessgen, Edward, and David Stargel. (2012, April). The digital twin paradigm for future NASA and US Air Force vehicles. In 53rd AIAA/ASME/ASCE/AHS/ASC structures, structural dynamics and materials conference 20th AIAA/ASME/AHS adaptive structures conference 14th AIAA (p. 1818).

4. Chen, Kang, et al. "Digital twins model and its updating method for heating, ventilation and air conditioning system using broad learning system algorithm." Energy 251 (2022): 124040.
5. Wang, Yujie, et al. "Digital twin and cloud-side-end collaboration for intelligent battery management system." Journal of Manufacturing Systems 62 (2022): 124–134.
6. Grieves, Michael. "Digital twin: Manufacturing excellence through virtual factory replication." White Paper 1.2014 (2014): 1–7.
7. Singh, Soumya, Max Weeber, and Kai-Peter Birke. "Advancing digital twin implementation: A toolbox for modelling and simulation." Procedia CIRP 99 (2021): 567–572.
8. Semeraro, Concetta, et al. "Digital twin paradigm: A systematic literature review." Computers in Industry 130 (2021): 103469.
9. Zhao, Guanjia, et al. "Hybrid modeling-based digital twin for performance optimization with flexible operation in the direct air-cooling power unit." Energy 254 (2022): 124492.
10. Steindl, Gernot, et al. "Generic digital twin architecture for industrial energy systems." Applied Sciences 10.24 (2020): 8903.

8

Real-World Case Studies of AI Integration with Thermal Energy Storage Systems

Naseer Ahmad Khan and Karem Elsayed Elfeky

8.1 Role of Artificial Intelligence in Thermal Energy Storage

Artificial Intelligence (AI), specifically machine learning (ML) and deep learning (DL), provides potent instruments to improve Thermal energy storage (TES) operations through predictive modeling, adaptive control, and autonomous decision-making. AI can evaluate extensive information from sensors, meteorological predictions, and energy consumption trends to enhance TES charging and discharging cycles, reduce energy waste, and integrate effectively with renewable energy sources such as solar and wind.

8.1.1 Case Studies of AI Integration with TES

Case studies collectively show how AI can completely change the performance and worth of TES systems. AI offers data-driven intelligence that makes energy management much better in many areas, from predictive repair in factories to dynamic load moving in district heating (DH) and smart buildings. AI is a key tool for the future of sustainable energy systems because it can predict demand, find the best charging and discharge times, change to outside factors, and work with digital twins (DTs).

8.1.2 AI Tools Used in the Case Studies

The case studies mainly used Artificial Neural Network (ANN), Feed-forward ANN, Nonlinear autoregressive ANN, least square support vector Machine (LS-SVM), and in some cases fuzzy LS-SVM and PID control integrated with AI.

- ANN was the most common because it is good at modeling complex nonlinear relationships in the TES system (predicting phase change material (PCM) behavior, thermocline thickness, or hot water tank performance).

DOI: 10.1201/9781003516132-8

- LS-SVM was used where fast computation was needed, even if slightly less accurate than ANN.
- Fuzzy LS-SVM added flexibility when the uncertainty in data (ground temperature) was high.
- PID + AI was applied for real-time water temperature control in TES using PCMs.
- In a larger, more complex setup (DT for concentrated solar power (CSP) molten salt storage), ML was combined with a physics-based model to enhance prediction and optimization.

8.1.3 Data Sources for AI Model

Data came from three main origins

- **Real system measurements** - sensor data from microgrids, district heating networks, geothermal wells, or CSP plants.
- **Experimental setups** – lab-scale TES prototypes, PCM heat storage tests, and water PCM hybrid systems.
- **Simulations** – computational models of TES behavior used to generate a training dataset where real data was limited.
- Additionally, weather forecasts and historical consumption patterns were key external inputs for demand prediction models.

8.2 AI Enhanced Energy Storage in Microgrids

Energy systems that are localized and can function independently or in conjunction with the primary grid are known as microgrids. There are substantial advantages to the integration of AI with energy storage in microgrids, including enhanced energy reliability, reduced costs, and more effective utilization of renewable energy sources. By predicting demand, managing supply fluctuations, and ensuring efficient energy distribution, AI is essential for optimizing microgrid energy storage.

In solar-powered microgrids, AI-enhanced TES optimizes energy storage by predicting solar generation patterns and adjusting charging and discharging cycles accordingly. This results in reduced energy wastage, improved grid stability, and cost savings. AI-driven algorithms analyze weather forecasts and real-time energy consumption data to determine the most efficient way to store and distribute thermal energy. In a case study from California, AI-managed TES led to a 30% reduction in energy wastage and enhanced grid stability with fewer power outages [1]. The integration of AI also ensured that excess solar energy was stored during the day and efficiently utilized during peak demand periods.

Industrial microgrids also benefit from AI-enhanced TES by managing waste heat and optimizing energy consumption. In manufacturing facilities, AI-driven TES systems analyze real-time temperature fluctuations to capture and store waste heat generated during production processes. This stored thermal energy is later used for heating applications, reducing the facility's dependence on external energy sources. A case study in Germany demonstrated a 25% reduction in energy costs and a significant reduction in carbon footprint through the implementation of AI-enhanced TES [2]. AI-powered sensors continuously monitored and optimized heat transfer processes, ensuring that energy was utilized efficiently with minimal losses.

Residential microgrids, such as smart communities, leverage AI-driven TES to enhance energy efficiency and sustainability. AI predicts household energy consumption patterns, optimizes storage cycles, and minimizes dependency on grid power during peak hours. In a Japanese smart community, AI-enhanced TES increased energy savings by 40%, ensuring a more reliable energy supply [3]. AI-controlled thermal storage systems efficiently manage excess solar energy, reducing peak load demand and providing cost-effective energy solutions to residents.

These case studies highlight the transformative potential of AI-enhanced energy storage in microgrids. By integrating AI-driven TES systems, microgrids can achieve greater energy independence, lower operational costs, and improved sustainability. The ability of AI to dynamically optimize energy storage and distribution ensures that microgrids can efficiently balance supply and demand, making them a critical component of future energy systems.

8.3 Application of AI in Geothermal Storage

Geothermal energy is a form of renewable energy derived from the Earth's internal heat. This heat originates from the radioactive decay of minerals and the residual heat from the planet's formation, creating a continuous thermal energy source beneath the Earth's surface. Geothermal energy can be harnessed for various applications, including electricity generation, direct heating, and TES. It is highly sustainable, low in emissions, and available 24/7, making it a reliable complement to variable renewable sources like solar and wind. The method of harnessing geothermal energy depends significantly on the depth and temperature of the heat source, leading to two main categories: shallow geothermal and deep geothermal systems.

Shallow geothermal, often referred to as **ground-source heat pump systems**, harnesses the relatively stable temperatures found less than 500 meters below the Earth's surface. These systems typically use horizontal or vertical closed-loop piping networks buried in the ground to extract or dissipate heat. They are most commonly used for residential or commercial heating and cooling applications. **Deep geothermal** systems tap into geothermal reservoirs located several hundred to several thousand meters below the Earth's surface. These systems are typically used for high-temperature applications such as electricity generation, industrial processes, or DH.

Figure 8.1 represents a closed-loop geothermal system. This system is designed to extract heat from deep underground and transfer it to a surface used for several applications, such as electricity generation or direct heating. The system consists of an inner and outer pipe through which heat transfer fluid (HTF) is circulated in a loop system within the well. Cold fluid is injected into the geothermal system through the outer pipe, descending to the deeper, hotter regions of the well. The HTF absorbs heat from the surroundings through conduction. The hot fluid ascends through the inner pipe due to the mechanical pumping. Once the hot fluid reaches the surface, its

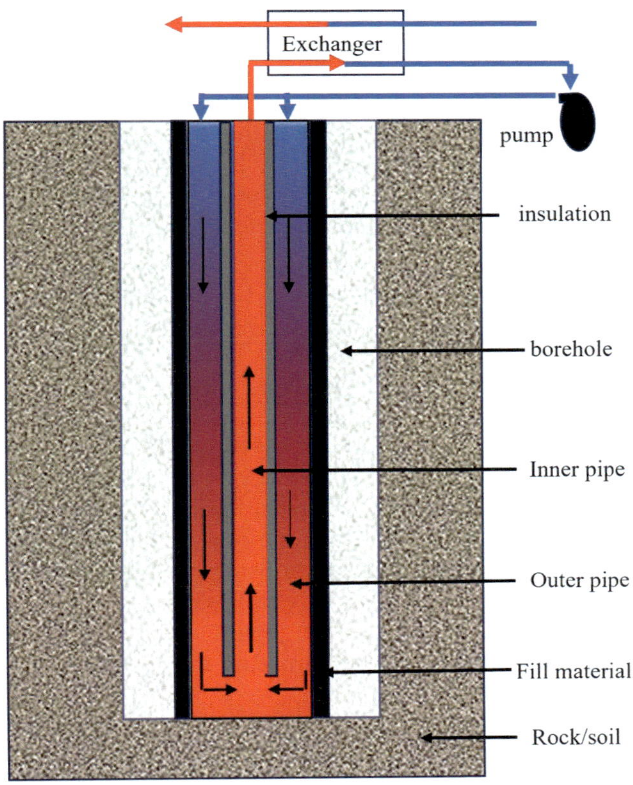

FIGURE 8.1
Deep borehole heat exchanger.

thermal energy is transferred to a secondary system, which could be a heat exchanger. After heat extraction, the cooled fluid is recirculated back into the well to repeat the process.

One significant real-world application of AI in geothermal TES is seen in geothermal DH systems. In countries like Iceland [4], where geothermal energy accounts for over 90% of heating needs, AI systems are being used to manage and forecast thermal demand across urban areas. ML algorithms analyze historical temperature data, user consumption patterns, and weather forecasts to predict future thermal needs. This enables utility providers to optimize the operation of geothermal plants and storage units, ensuring that heat is generated and stored during off-peak periods and efficiently distributed when demand rises. These AI-driven forecasting models significantly reduce energy waste, improve operational efficiency, and enhance the reliability of heat delivery systems.

In Canada, AI-enhanced geothermal TES is being deployed in residential and community-scale applications through ground-source heat pump systems. These systems use underground loops to exchange heat with the Earth. AI algorithms control the timing and intensity of heat exchange based on indoor occupancy patterns, external temperatures, and electricity tariffs. During low-tariff periods or times of excess renewable energy generation, the system stores thermal energy in the ground. During high-demand or high-tariff periods, the stored energy is utilized to meet heating or cooling needs. This intelligent control significantly improves energy efficiency and reduces electricity bills for end-users. Moreover, AI models continue to learn and adapt, further refining the system's performance over time.

8.4 AI Integrated to Solar Energy Storage

Optimizing solar energy storage is an essential element in the transition towards sustainable and renewable energy systems. Solar energy, while abundant and clean, is inherently intermittent due to its dependency on sunlight availability, which varies by time of day, weather conditions, and seasons. As such, effective storage solutions are required to ensure that the energy generated during sunny periods can be stored and used during times when sunlight is not available, such as nighttime or cloudy days. Optimizing these storage systems is critical to maximizing the efficiency, reliability, and economic viability of solar power.

There are several types of storage systems commonly used for solar energy, including chemical (batteries), thermal, and mechanical storage systems. Lithium-ion batteries are currently the most popular for both residential and commercial solar applications due to their high energy density, efficiency, and declining costs. However, other battery types, such as flow batteries, lead-acid, and sodium-sulfur, also present unique advantages depending on specific application requirements. TES, which involves storing solar energy in the form of heat using materials like molten salts or PCMs, is particularly useful for CSP systems. Mechanical storage, such as pumped hydro or compressed air, is typically used for grid-scale applications and requires specific geographic conditions.

Optimization involves enhancing the efficiency of energy conversion, minimizing energy losses, and managing the charging and discharging cycles. Energy Management Systems (EMS) play a crucial role in this regard. These systems use algorithms to analyze energy production and consumption patterns, ensuring that energy is stored when there is surplus and retrieved when demand peaks. AI and ML are increasingly being integrated into EMS to improve predictive capabilities, enabling smarter decision-making and real-time optimization of storage operations. For instance, AI can predict energy generation based on weather forecasts and adjust storage parameters accordingly.

Figure 8.2 represents a solar thermal energy system with an intelligent control mechanism that directs the flow of HTF through a closed-loop circuit. The process begins at the solar collector, which absorbs solar radiation and converts it into thermal energy. The HTF circulates through the collector, gaining heat as it moves. The control algorithm continuously monitors real-time data such as solar irradiance, fluid temperature, ambient temperature, and demand levels. Based on this data, the algorithm determines when to activate the circulation pump, ensuring the HTF only flows when sufficient solar energy is available, thereby maximizing system efficiency and preventing energy losses.

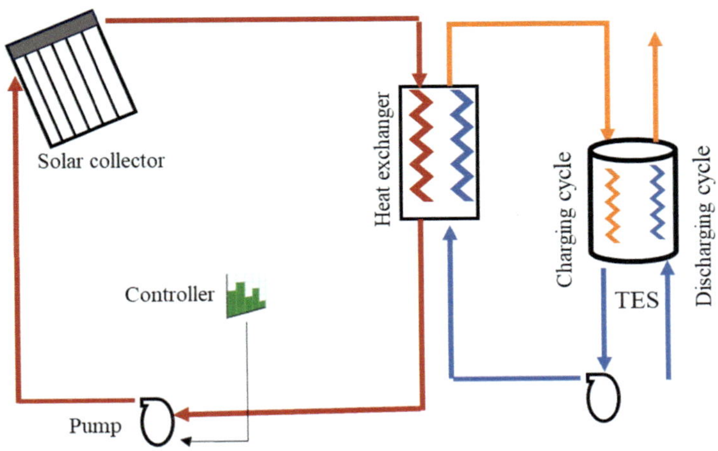

FIGURE 8.2
Solar heating system integrates with TES.

Once heated, the HTF flows from the solar collector to a heat exchanger, where it transfers thermal energy to another loop. This second loop is connected to a TES tank. In the TES tank, the transferred heat is stored for later use, allowing the system to decouple energy collection from consumption. The TES system enables time-shifting of energy use, meaning that solar energy collected during the day can be used during the night or periods of low solar availability. A separate circulation pump manages the heat delivery from the TES to the end-use (e.g., space heating or domestic hot water), again under the guidance of the control algorithm.

To optimize solar energy storage at a broader scale, policy and regulatory frameworks play a vital role. Clear policies, standardized interconnection procedures, and supportive regulations can remove barriers to storage deployment. Utility companies and grid operators must also adapt to accommodate decentralized and distributed energy storage systems. Collaboration between governments, industries, research institutions, and communities is necessary to drive innovation and develop best practices.

8.5 Digital Twin Solution for Medium to High Temperature TES

DT is an interactive digital model of a physical system that replicates its functionality, state of health, and operation in real time. DTs provide robust options for TES system simulation, monitoring, control, and optimization by combining physics-based models with real-time data analytics and ML. Traditional models fail to capture the complicated interactions present in medium to high temperature TES systems due to the dynamic and complex nature of thermal gradients, material strains, and energy fluxes. The DT can help with this. In a CSP plant's molten salt TES system, for instance, the DT can model phase change behavior, heat transfer dynamics, and thermomechanical stresses on containment vessels; this information is useful for optimizing performance and reducing risk. Medium to high temperature TES systems, which typically operate in the range of 100–600°C or higher, are particularly important for industrial processes, CSP plants, and DH networks.

One of the most transformative aspects of DT solutions is their integration with AI and ML algorithms. These technologies enhance the predictive capabilities of the DT by identifying patterns, correlations, and anomalies that may not be apparent in conventional models. For example, ML can be used to fine-tune model parameters based on historical performance data, improving the accuracy of predictions over time. Reinforcement learning algorithms can be employed to develop optimal control policies that adapt to evolving operational contexts.

Moreover, AI-driven DT can operate autonomously, making real-time decisions without human intervention. This is particularly useful in remote or large-scale TES installations where manual monitoring and control would be impractical. AI can also help in scenario analysis and strategic planning by evaluating the impacts of different market conditions, regulatory changes, or technological upgrades.

8.6 AI-Driven Heat Storage in DH Systems

DH system is the process in which thermal energy from a central system is supplied to domestic households or commercial buildings to provide space heating or to fulfill their hot water needs. This technology is growing country-wise and putting its fair share in the heat energy generation map [5]. The goal is not just to produce the heat energy to fulfill the heat demand but also to reduce the carbon emissions that are becoming a huge reason for environmental changes [6], burning of fossil fuels is the major affecting factor in this regard. Space heating and hot water needs for winter have a huge share in the energy demand of around 70% overall, resulting in a high impact on environmental changes by this specific sector. With time and technological improvements, the method for DH is evolving and is in its 4th generation, as shown in Table 8.1.

First generation was based upon fulfilling the DH requirements by steam boilers which used coal as a fuel, second generation involved combined heat & power (CHP) plants and steam boilers utilizing coal and oil both as fuels, shown in Figure 8.3. third generation had large-scale CHP plants but with fuel improvements including biomass, waste, and conventional resources. Today, the 4th generation DH systems are utilizing renewable energy resources for providing the heat demands.

Another configuration is a DH system using renewable energy resources and incorporated with seasonal thermal storage. A combination of solar energy and seasonal thermal storage is common for DH systems. Seasonal TESs (STES) are used to store the amount of excess energy in the hours when heat demand is less, and energy production is high, same stored energy is then utilized in the peak load hour to meet the heating demand, resulting in

TABLE 8.1

DH Technology Generation

Generation	Heat Production Method	Fuel Type	Time Span
1st generation	Steam boilers	Coal	1880–1930
2nd generation	CHP & Steam Boiler	Coal & oil	1930–1980
3rd generation	Large-scale CHP	Biomass, waste, and Fossil fuel	1980–2020
4th generation	Heat recycling	Renewable energy, Waste heat recovery	2020–2050

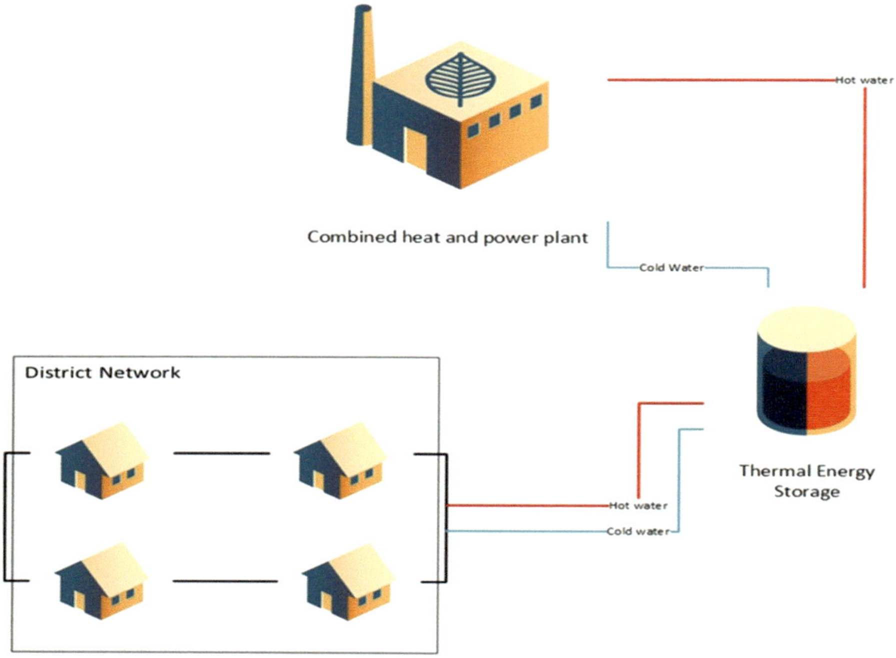

FIGURE 8.3
Model of DH system.

less fuel consumption and reduced gap between supply and demand cycle. STES technologies have a major use for the storage of thermal energy from one season to use it in a different season. STES has different configurations and types for storing thermal energy and then supplying it to domestic and commercial entities through DH and cooling setups [7]. Sensible heat storage is the best option among all other types of TESs to obtain better solar fraction for space heating [8]. The best-known types of seasonal TES variants are aquifer storages, borehole storages, pit storages, and seasonal tank storages.

Aquifer and borehole storage systems utilize the geothermal environment, i.e., natural ground, to store the thermal energy and utilize it during the demand period [9]. The main parameters for determining the heat transfer and losses for the storage are thermal properties of the storage medium, time of storage, storage temperature, storage geometry, and volume. Due to its low cost, as well as its ability to deal with the large time scales, ground storage is the most promising technology for community and DH needs [10]. Figure 8.4 represents different types of STES.

Figure 8.5 illustrates an integrated solar-assisted DH system that combines short-term thermal storage (STTS) and long-term thermal storage (LTTS), primarily through a borehole storage system, and is managed by a central

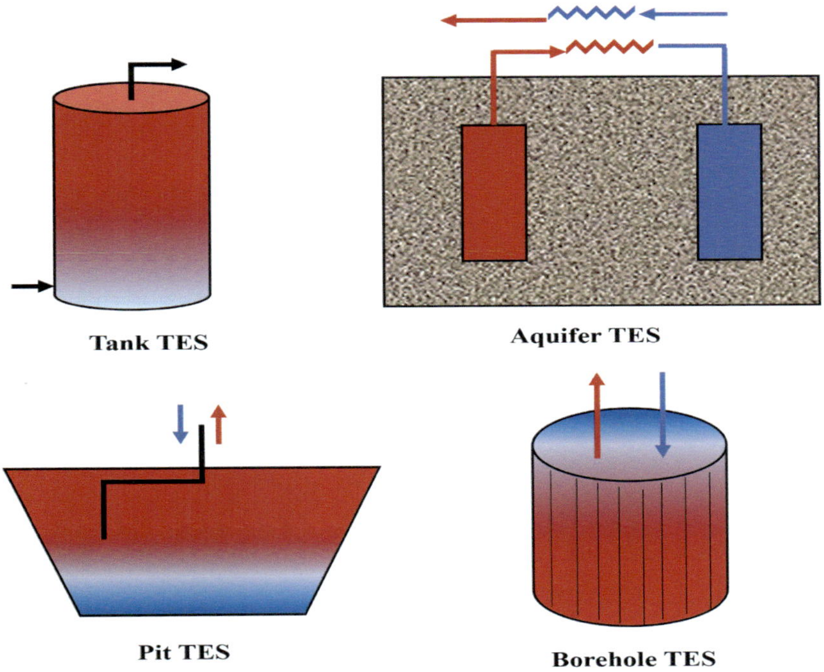

FIGURE 8.4
Different types of seasonal TES.

energy center with the help of a sophisticated controller system. The system is designed to supply heating demand reliably and efficiently by maximizing the use of renewable solar energy while providing backup support through an auxiliary boiler. The role of the controller is crucial; it orchestrates the heat flows between various subsystems, optimizes charging and discharging cycles, and ensures that the energy demand of the district network is met sustainably and cost-effectively.

Starting from the solar field, thermal energy is harvested using solar collectors and circulated through the solar loop. The HTF carries this energy to a heat exchanger, which isolates the solar loop from the STTS loop, thus allowing the solar energy to be transferred without mixing the working fluids. The heated fluid from the solar side is used to heat water in the STTS loop, which then carries the energy to the energy center. The energy center, consisting of short-term thermal storage tanks, acts as the main hub where thermal energy is temporarily stored and redistributed as needed.

From the energy center, heat can follow three main paths depending on system demand, solar availability, and controller decisions. First, it can be sent directly to the DH network via another heat exchanger that connects the STTS loop to the district loop, providing immediate heating to residential or

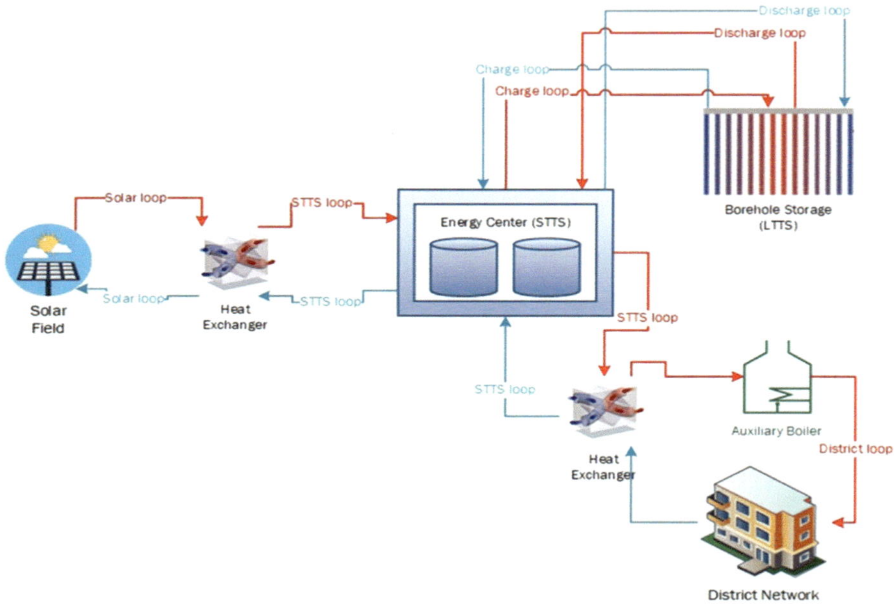

FIGURE 8.5
DH network model of the proposed model.

commercial buildings. Second, in periods of excess solar heat (typically during summer), the controller can activate the charge loop, directing thermal energy to the borehole TES system for seasonal storage. This stored energy can be later retrieved during colder months using the discharge loop, sending heat back to the STTS tanks for redistribution.

Third, when solar energy and stored heat are insufficient, particularly during high-demand periods or extended cloudy conditions, the auxiliary boiler is triggered. The boiler is supplied by the STTS loop and discharges into the district loop to ensure uninterrupted heat delivery. This hierarchical energy distribution strategy, managed by the controller, prioritizes renewable sources and reserves fossil fuel backup only for critical demand scenarios.

The controller is the intelligence behind this integrated network. It continuously monitors key system variables such as temperature, pressure, flow rates, solar radiation, and user demand. Based on real-time data and predictive algorithms (which may incorporate weather forecasts, historical usage patterns, and energy pricing), the controller dynamically decides when to charge or discharge the LTTS, when to route solar energy directly to the district network, and when to engage the auxiliary boiler. Its objectives are to maximize solar energy utilization, minimize operational costs and emissions, and ensure thermal comfort for end-users. Additionally, the controller can include fault detection and diagnostics, alerting operators to maintenance needs or system anomalies before they escalate.

Heat storage within DH involves capturing excess thermal energy during periods of low demand and storing it for use during peak periods. This process enhances the system's ability to balance supply and demand, integrate renewable energy sources, and improve overall efficiency. AI enhances this process by employing advanced algorithms and ML techniques to predict demand, optimize storage and distribution, and manage resources dynamically.

Prior research on TES prediction using a variety of AI methodologies is summarized in Table 8.2. A variety of AI algorithms, such as ANN, are employed in these investigations, which incorporate a variety of TES systems. Although the majority of research is devoted to the prediction of TES system performance, a few also concentrate on the optimization of system design. AI models were trained using a variety of data sources, including experimental, simulation, and weather data.

TABLE 8.2

AI Techniques for Control and Optimization of TES

TES	AI Method	The Objective	Main Finding	Ref.
Multichannel thermal storage tank	ANN	Predicting the liquid fraction	The average temperature and liquid fraction of PCM	[11]
System for thermal storage that employs water-PCM	ANN	Predicting the solar fraction of the hybrid solar thermal storage system that incorporates PCM	A remarkable R2 value of up to 0.9999 was attained by the employed ANN.	[12]
PCM composite Structured with a metal foam-fin for hot water storage.	ANN	PCM phase transition forecasting	Excellent prediction performance was demonstrated by the ANN.	[13]
Hot water storage tank	Feed-forward ANN	The hot water storage reservoir design of the solar water heating system is being verified.	The maximal error of only 2.62% was observed in the overall conductance predictions of both the mathematical and ANN models.	[14]
Temperature of externally melting ice TES	Feed-forward ANN	Enhancing the efficiency of the large temperature differential refrigeration system	Despite the presence of varying thermal loads, the neural network controller model maintained its effectiveness.	[15]

(Continued)

TABLE 8.2 (*Continued*)

AI Techniques for Control and Optimization of TES

TES	AI Method	The Objective	Main Finding	Ref.
Borehole	Neural network and LS-VM	Assessing the ground temperatures of two boreholes.	Although the prediction accuracy of the fuzzy LS-SVM was somewhat inferior to that of the ANN model, it exhibited a much quicker performance.	[16]
Stratified thermal storage tank	ANN feed forward	Predicting the thermocline thickness and half-cycle figure of merit of a layered thermoelectric storage tank in a gas district cooling system	Better prediction performance and a very high R2 value	[17]
PCM	PID control	Regulating the water heating temperature in thermal storage using phase change and valley electricity.	Throughout the thermal storage process, the temperature of the heating water stabilized at 57 °C, with a deviation of 2 °C, indicating satisfactory heating efficiency.	[18]
Residential hot water reservoir	Nonlinear autoregressive ANN	Optimization of residential hot water output in a fifth-generation DH and cooling energy transmission system	The entire optimization was achieved by relocating 14% of the power consumption from on-peak to off-peak hours.	[19]

8.7 Challenges when Using AI with TES and Solutions

Data quality and availability – addressed by combining simulated and experimental data and using real-time sensor feeds.

Nonlinear and time-dependent behavior of TES – handled using ANN and advanced ML algorithms that can capture temporal patterns.

Integration with control systems – solved through predictive controls (ANN-based model predictive control in DH) and by AI in digital twins for real-time decision making.

Climate variability – addressed through continuous model retraining using updated local weather and demand data.

8.8 Practical Adoption by Government/ Companies and Main Barrier

Government and companies could use AI-TES to reduce fuel cost and carbon emissions in DH, improve reliability in renewable-powered microgrids, and optimize industrial waste heat recovery.

Barrier

- High initial investment in sensors, control infrastructure, and skilled workforce.
- Regulatory gaps in integrating AI- based controls into critical energy systems.
- Data security and interoperability issues when connecting AI systems to existing energy networks.

In countries with different climates (humid tropic, cold desert) or weaker infrastructure, the AI model would need Retraining with local weather and demand data. Possible simplification of algorithms, a low-cost sensor to replace high-end industrial monitoring systems. In places like Iceland (abundant geothermal) or California (strong solar resource), AI models are trained on abundant high-quality local data.

References

1. Al Marzooqi, M., *The Contribution of Artificial Intelligence (AI) to Disaster Response and Management*. 2024, Anglia Ruskin Research Online (ARRO).
2. Miehe, R., et al. *Regional carbon footprints of households: A German case study*. Environment, Development and Sustainability, 2016. **18**: p. 577–591.
3. Ukoba, K., et al. *Optimizing renewable energy systems through artificial intelligence: Review and future prospects*. Energy & Environment, 2024. **35**(7): p. 3833–3879.
4. Meirbekova, R., et al. *Opportunities and challenges of geothermal energy: A comparative analysis of three European cases—Belgium, Iceland, and Italy*. Energies, 2024. **17**(16): p. 4134.
5. Rafiee, A., and K.R. Khalilpour, *Chapter 11 – Renewable Hybridization of Oil and Gas Supply Chains*, in *Polygeneration With Polystorage for Chemical and Energy Hubs*, K.R. Khalilpour, Editor. 2019, Academic Press. p. 331–372.
6. Bint Faheem, J.J.I. *And engineering*. Energy Crisis in Pakistan, 2016. **3**(1): p. 1–16.
7. Kalaiselvam, S., and R. Parameshwaran, *Chapter 7 – Seasonal Thermal Energy Storage*, in *Thermal Energy Storage Technologies for Sustainability*, S. Kalaiselvam and R. Parameshwaran, Editors. 2014, Academic Press: Boston. p. 145–162.
8. Xu, J., R.Z. Wang, and Y. Li. *A review of available technologies for seasonal thermal energy storage*. Solar Energy, 2014. **103**: p. 610–638.

9. Bott, C., et al., *State-of-technology review of water-based closed seasonal thermal energy storage systems.* Renewable and Sustainable Energy Reviews, 2019. **113**: p. 109241.
10. Rad, F.M., A.S. Fung, and M.A.J.E. Rosen, *An integrated model for designing a solar community heating system with borehole thermal storage.* Energy for Sustainable Development, 2017. **36**: p. 6–15.
11. Xiao, T., et al. *LSTM-BP neural network analysis on solid-liquid phase change in a multi-channel thermal storage tank.* Engineering Analysis with Boundary Elements, 2023. **146**: p. 226–240.
12. Eldokaishi, A., et al. *Modeling of water-PCM solar thermal storage system for domestic hot water application using artificial neural networks.* Applied Thermal Engineering, 2022. **204**: p. 118009.
13. Cui, W., et al. *Heat transfer analysis of phase change material composited with metal foam-fin hybrid structure in inclination container by numerical simulation and artificial neural network.* Energy Reports, 2022. **8**: p. 10203–10218.
14. Kulkarni, M.V., D. Deshmukh, and S. Shekhawat. *An innovative design approach of hot water storage tank for solar water heating system using artificial neural network.* Materials Today: Proceedings, 2021. **46**: p. 5400–5405.
15. Dong, W., L. Kun, and W. Yi. *Model Predictive Control of a Large Temperature Difference Refrigerating Station with Ice Cold Thermal Energy Storage,* in *2017 Chinese Automation Congress (CAC).* 2017, IEEE.
16. Zhou, S., et al. *Prediction of the ground temperature with ANN, LS-SVM and fuzzy LS-SVM for GSHP application.* Geothermics, 2020. **84**: p. 101757.
17. Amin, M., A. Majid, A.A. Soomro, and A. Akmar. *Artificial neural network modelling approach for assessment of stratified thermal energy storage tank,* 2026. **14**, no. 19: pp. 3371–3378.
18. Yu, H., C. Yang, and Q. Zhao. *Division of Working Variable Domain Fuzzy PID Control for Valley Electric Phase Change Heat Storage,* in *2019 IEEE Innovative Smart Grid Technologies-Asia (ISGT Asia).* 2019, IEEE.
19. Buffa, S., et al. *Fifth-generation district heating and cooling substations: Demand response with artificial neural network-based model predictive control.* Energies, 2020. **13**(17): p. 4339.

9

Future Trends and Challenges in AI-Powered Energy Management Systems

M. Asim Amin, Muhammad Waseem, and Youwei Jia

9.1 Introduction: Role of AI Adoption in EMS

The worldwide adoption of RES has been at the forefront. The global adoption of RES capacity rose to 3372 GW by the end of 2022, with a 9.6% increase in a single year, reaching a capacity of 295 GW. It is important to note that the newly installed capacity from RES in 2022 is from solar, wind, and other clean energy resources. The fast adoption of RES is fundamental to energy transition and paves the way for data-centric grid management, enabling the maximization of RES usage alongside its expansion. Many solution approaches exist that discuss and help overcome the RES intermittency and its operational challenges using AI [1].

On the other hand, DeepMind implemented AI-based approaches for wind farm operations by accurately forecasting the wind power, boosting their economic value to ~20% and their optimal scheduling, and providing support in fulfilling profitable demand and supply balance and improving grid stability [2]. It is worth noting that energy corporations use AI-driven solutions to forecast demand, grid management, and predictive maintenance optimally. Google hired DeepMind to optimally manage and reduce cooling energy demand by 40% [3], further highlighting the adoption of AI. It can be concluded that the impact of AI and its adoption for forecasting and EMS is exploding. In addition, the projected growth of RES and AI adoption in EMS over the years from 2010 to 2030 is presented in Figure 9.1, highlighting the AI adoption and rise in the use of RES.

9.2 Emerging Trends in AI

With technological advancements, AI-driven solutions for EMS are becoming more promising and effective. The possible key areas are:

DOI: 10.1201/9781003516132-9

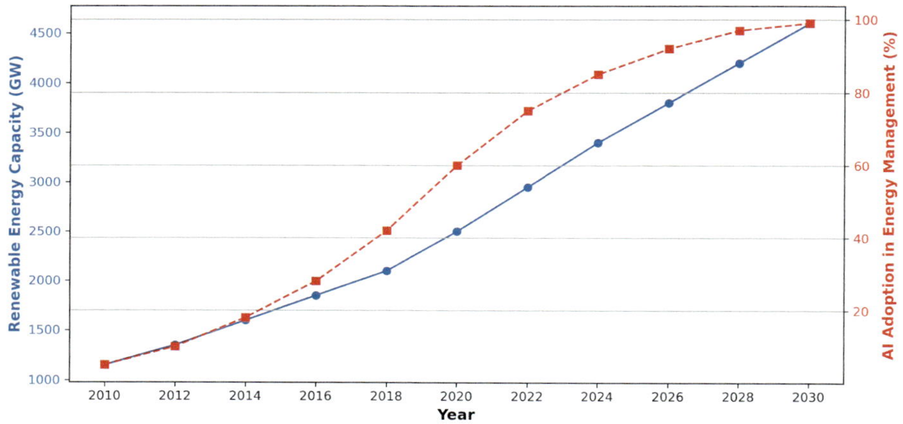

FIGURE 9.1
Projected growth of RES for AI adoption in EMS (Expert forecast of EMS from BCG and IBM [4, 5], RES [6, 7]).

9.2.1 Big Data

The modern system generates tremendous data from the electric sensors, meters, and connected devices. Some terms deal with extensive data, with a high amount of data to analyse and get valuable insights. It is important to note that big data may play a vital role after processing the data from the metrological system and other energy system data to improve the system performance, operational efficiency, and customer satisfaction by boosting overall social welfare [8]. On the other hand, the system operator on the grid level enhances accurate decision-making by analysing historical data and guaranteeing precise forecasting and identification of anomalies. However, big data requires complex and innovative solutions to analyse the data to further decide on optimal control.

9.2.2 Machine Learning

Advanced Machine Learning (ML) based methods are essential for the energy sector and are used to forecast intricate behaviours for energy consumption, RES, and overall performance. On the other hand, the conventional ML approaches, advanced deep reinforcement learning, large language models, and decision trees have been developed for energy applications such as short and long-term forecasting [9], fault predictions in equipment, etc. It is worth noting that the deep learning models have outperformed conventional methods in more accurate forecasting. Due to these precise results, the grid operator can maximize and actively manage the supply and demand. In comparison, reinforcement learning-based methods are being investigated in more detail for their real-time implementations of battery allocation, resource management, electric vehicle scheduling for cost minimization [10], and grid stability.

9.2.3 Internet of Things

IoT implementation in the EMS offers real-time data monitoring flexibility through smart devices. In a smart grid, several sensors and smart meters provide stable monitoring and connectivity in both directions, enabling massive data for decision-making [8]. It further allows the AI-based model to make real-time decisions and the data accessed through the IoT devices to offer required changes, such as storage, energy sharing, trading, etc., and improve the grid performance and real-time control/monitoring of energy distribution. It is important to note that combining big data and IoT provides a framework that ML uses to maximize energy usage.

Overall, big data and IoT are the fundamental components to achieve optimal resource management through the sensors on the consumer and grid level (through the phase measuring units and smart meters) to analyse the load forecasting control and thereby improve grid stability and operational productivity. In addition, the IoT-enabled ecosystem shown in Figure 9.2 and the advanced structure of the power grid improve the real-time management of end-user demand through the IoT-based sensors and ML and big data analysis. On a larger scale, there is a need for a framework to handle

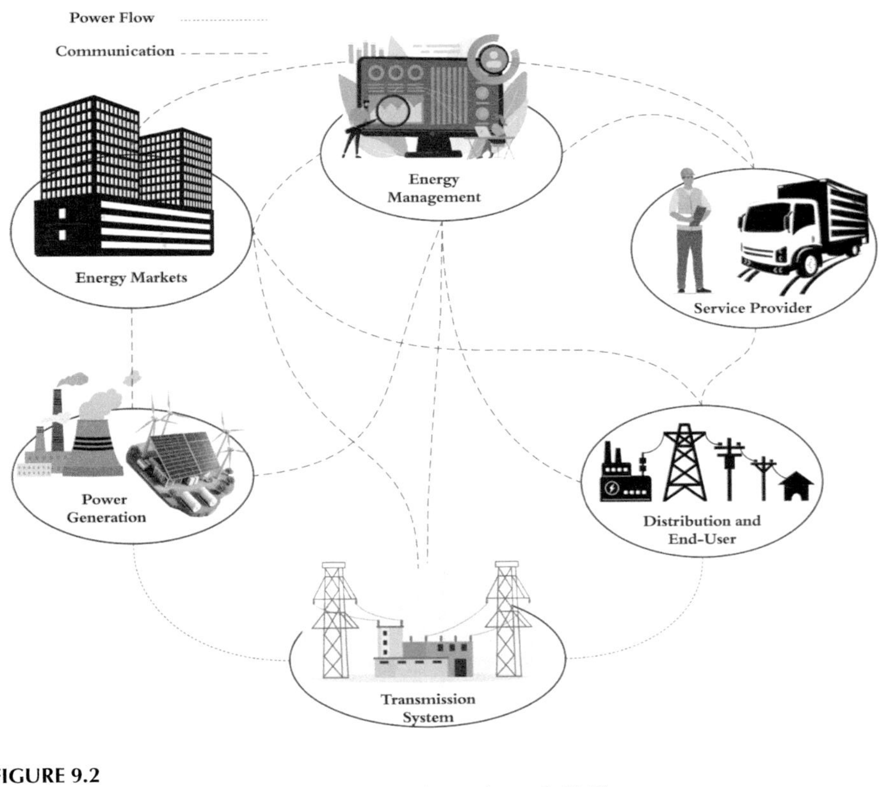

FIGURE 9.2
IoT-driven ecosystem to maintain demand balance through EMS.

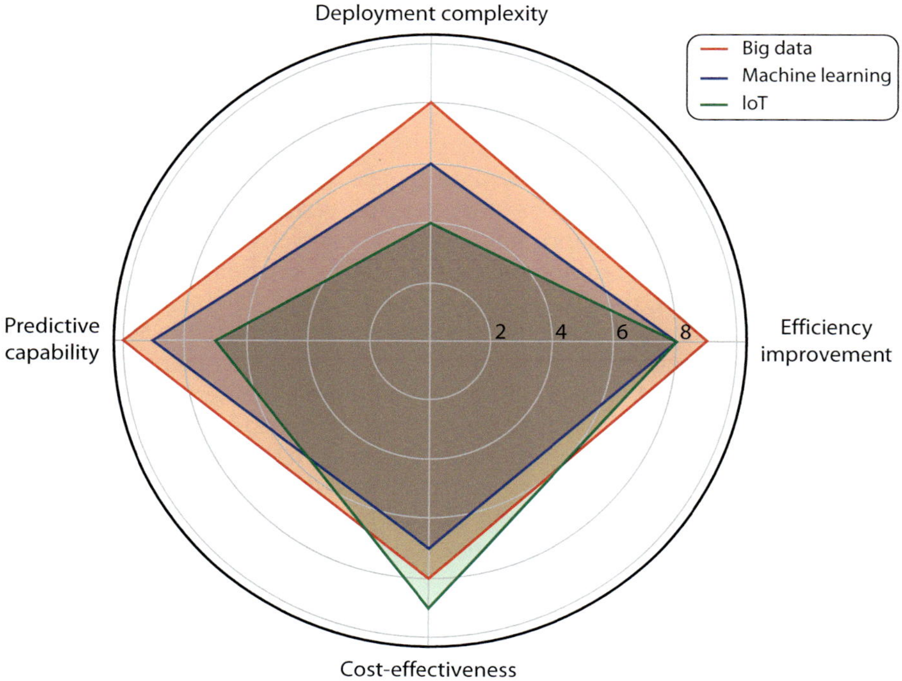

FIGURE 9.3
Comparative analysis of emerging trends in energy efficiency (1–10 range based on complexity) [8].

extensive data and guarantee privacy and security in an IoT-based system. In addition, comparing the emerging trends helps improve the overall EMS. ML possesses remarkable prediction skills and improves forecasting precision despite adding complexity to its application. IoT is adequate for real-time control and monitoring by offering feasible solutions, but it depends on big data and ML for decision-making. In addition, big data provides a necessary framework to manage the data generated through the IoT. Hence, an amalgamation of these advancements fosters efficient EMS that surpasses their singular capabilities and comparison given in Figure 9.3.

9.3 AI-Driven EMS and Storage

Based on recent developments, AI has a wide range of novelties, including storage, transmission, and energy balance. However, the promising efforts are as follows:

9.3.1 Predictive Maintenance

It enables forecasting of equipment failure before it happens. *"Predictive Maintenance"* of electrical systems through the availability of data and the implementation of AI approaches to make predictive decisions are the prominent uses of AI. In RES, wind turbines and solar farms are energy assets with excessive data availability, such as temperature, generation, and vibrations, and they search for prior decisions regarding the failure and degradation of the component [1]. It enables the operator to react quickly to foresee the equipment problem and helps reduce unplanned downtime and avoid major breakdowns. Forecasting a gearbox fault for a wind turbine on a week-ahead basis can help prevent an unanticipated failure and facilitate scheduled repairs. Several studies show that the AI-based maintenance solution helps to reduce downtime, lower maintenance costs, and enhance the component lifetime. Optimally planning the operation and maintenance duration increases the system's reliability and helps RES be financially feasible.

9.3.2 Real-Time Management

The deployment of EMS and storage systems and flexible load management is essential for timely decision-making regarding battery charging/discharging and further distribution in the electric grid. AI-enabled smart grids can dynamically adjust demand and supply and monitor and control them in real-time, thereby improving the utilisation of distributed resources. It includes further optimising the battery performance concerning the solar or wind power output change and reducing the power flow to avoid line congestion [11]. Based on the existing approaches, the Deep Reinforcement Learning (DRL) based agents create a more realistic resource utilisation by addressing the storage, generation, and scenario-based solutions. It can improve overall efficiency, avoid overloads and RES curtailment, and allow price arbitration to lower overall energy costs by utilising AI-based forecasting and grid stability monitoring. With these advancements, the more autonomous grid and the storage system can be depicted by constantly optimizing the energy distribution by managing the consumption patterns and lowering peak demands.

9.3.3 Digital Twin

It is a virtual representation of an actual system that operates concurrently to emulate its action, becoming a formidable instrument for research and real-time management in the energy domain. Integrating digital twins with IoT enables the model to present an innovative method for enhancing the control system and its long-term viability and operational effectiveness. It further allows the exploration of the optimization of real-time scenarios within the virtual model before implementing it in the real-time model. Recent developments emphasize applying digital twin models to support predictive maintenance, grid model improvements, and control systems. For example, a wind farm can replicate the physical farm with a digital twin model based on AI to forecast a wind farm's impact on weather incidents and change the setting accordingly [12]. It can further enable the integration of the heating and electrical systems to make a multi-energy system for experimentation and identification of accurate control methods. Hence, digital copy integration allows control and monitoring of the physical system.

Hence, the rapid development in energy systems through AI-based models improves resource trustworthiness, enables adaptable load allocation, improves forecasting precision, and further enables the self-sustaining and strong energy system. In the transformation, integration must assure operational effectiveness and price reduction.

9.4 Challenges in the Energy System

The use of AI in EMS unveils the considerable challenges that require solutions that are yet to be sought and discussed in the following section.

9.4.1 Security and Privacy

The energy system is a critical part of the necessary infrastructure. Thus, the safety and protection of information are crucial to the system. The vast amount of data used by AI consists of data from the different sensors and energy meters, raising questions about safety and confidentiality [13]. Utility companies limit the availability of operational data and grid code information due to security precautions, preventing data availability for AI-based research [14]. It is also essential to consider that access to data may lack uniformity and accuracy, as different companies might not record all the data, events, or faulty sensors. Hence, trustworthy, high-quality data is lacking, hindering the training of AI-based models and preventing their use in the industry [1]. The data availability may also be impacted as the AI solution for one grid is unsuitable for another. Moreover, the AI models might generate new shortcomings, such as malicious attacks on the model, making AI-based tools pose security concerns as a top priority. Hence, it can be depicted that AI implementation in the energy sector is challenging without the availability of data accuracy, management, safety regulations, and compatibility criteria.

9.4.2 Barriers and Policies

Due to existing barriers to AI implementation through institutional and human intervention, the main obstacle is the required skill set, as deploying advanced AI algorithm implementation requires expertise that traditional companies may not have. There is a need for more data scientists, experts, and engineers with strong skills, which are required by industries to train the staff and hire experts [15]. In addition, the next issue is trust in AI and organizational culture, as the operators are not confident using the *"black-box"* AI models, with an opaque performance over the decision-making process. It raises ethical issues regarding accessibility, administration, and fairness. For example, stakeholders must guarantee that AI-based actions are reliable and fair to prioritize power restoration in the priority areas. The governments are working on making the legislation fit with AI development, as the recent laws regarding energy lack autonomous decision systems or data exchange settlements [16]. It is important to note that the control of AI in the energy sector critically depends on a code of conduct, industry standards, and formal law-making. The industry needs to guarantee and execute staff members who can effectively manage and oversee AI-based devices. Ignoring the government and socio-economic policies may slow the deployment of AI or deliver more suboptimal results.

Despite its disruptive potential, implementing AI in energy systems faces various challenges, including privacy concerns, regulatory uncertainty, and an absence of skilled professionals. Moreover, the deficit of well-defined regulations and the environmental impact of AI computation systems raise future sustainability problems. Dealing with these challenges requires coordinated efforts in the energy sector, regulations, and knowledge building to ensure secure and efficient use.

9.5 Potential Solutions

Therefore, the stakeholders are pursuing numerous strategies to overcome the challenges in an energy system.

9.5.1 Open Science and Data Sharing

Classifying the engagement in data sharing is an essential approach through collaboration between academics, businesses, and the government sector, and it is crucial to develop a data-sharing platform. Developing consistent guidelines and norms for data types, compatibility, and effectiveness is also important. Utilities and academic institutions can collaborate to train the model on the available open data and train the model on publicly shared data. The solution required protocols that augment privacy, which are critical for these approaches. On the other hand, federated learning shows an approach in which the AI model trains from the local data and stores it in centralized storage based on their sensitivity [14]. It enables the attainment of great insights while maintaining privacy. There is a need for frameworks or standards to enhance data access in a very controlled way by maintaining confidentiality and handling challenges to create a more flexible AI-based EMS system.

9.5.2 Regulations and Policies Framework

Fostering and guiding AI usage usually depends on the government, and legal legislation is required. Under some scenarios, the government may provide subsidies and assist in the research and development of AI-based solutions to make the grid more autonomous and provide grid stability [17]. There is a need to develop AI guidelines similar to the current grid standards and offer a rigorous framework for different businesses. In addition, connecting to EU AI Act conformity frameworks ensures the AI follows a risk control and transparency guideline. According to stakeholders, the augmentation of law might help reduce the associated challenges and increase the benefits by integrating AI in market conditions and grid control areas [13]. Founding a strong legal framework and a sandbox that can enable the higher authorities to consider investing in AI is vital.

9.6 Background and Future Prospective

In development over the years, the advancement from the rule-based to the modern AI model until 2022 and its further improvements, as shown in Figure 9.4, highlights the changes. The initial effort between 1950 and 1970 was to automate an expert system related to the grid and the start of fuzzy logic and neural networks in 1980 [18]. On the other hand, the AI implementation for energy problems started in 1990, and further, the demand response-based problem utilised AI from 2000. In (2011–2014), IBM Watson and DeepMind collaborated on building energy usage and data centre productivity. In addition, AI provides a promising solution for digital twin-based approaches for autonomous EMS and optimizes it further in real-time. It can be depicted from the availability of IoT-based data, and AI-based predictive management can enable the target of independent and self-healing with minimal human intervention. The available data from IoT supports advancement in big data for data processing and develops efficient algorithms for driving real-time decision-making, which was further paired in 2010 to support the modern grid with AI-based forecasting, trading, and control established by 2023.

Until now, advancements in this field have provided practical innovations in forecasting control systems by utilizing advanced generative models to enable real-time decision-making and improve efficiency. In addition, the privacy problems of federated learning play an important role, as [19] shows

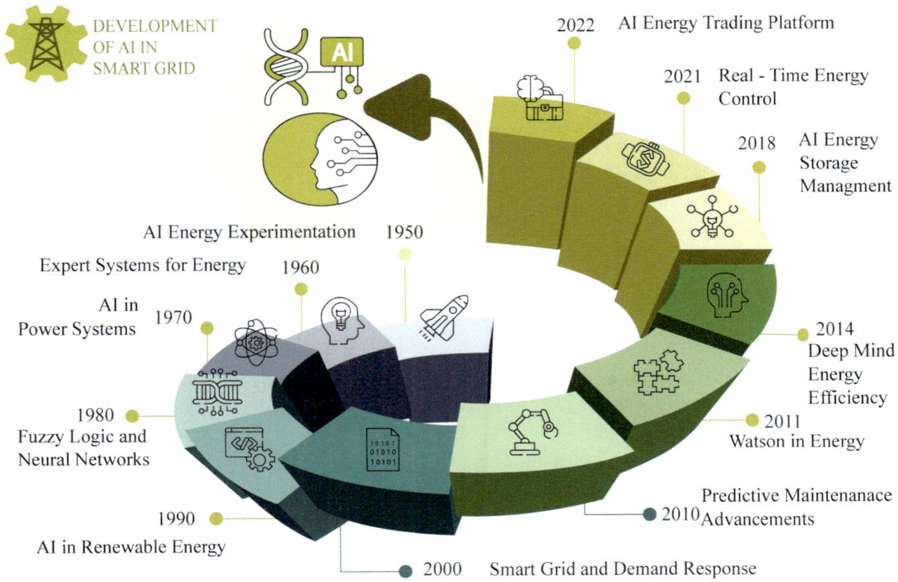

FIGURE 9.4
Development of AI for smart grid.

that load forecasting is trained over the smart meter datasets without revealing the end-user information. Likewise, integrating generative AI with multiagent control and the digital-twin of power systems can replicate the grid dynamics and cooperate further to maximize the grid operation and overcome the challenges associated with Distributed Energy Resources (DERs) [20]. Further advancements are set up for the autonomous grid, and new standards, IEEE-1547.9 [21], are being set up for the insertion of battery storage at the system level, with the grid forming inverter technologies by controlling the voltages and frequency locally. In addition, the European Union AI Act, finalized in 2024, requires the *"AI Act and Regulatory Sandboxes"* to access autonomous real-life advancements [22].

Several Asian countries, including Singapore, have initiated the test of virtual power plants using AI for an adaptable grid resource by combining solar, battery, and EV charging stations. On the other hand, the US Department of Energy has allocated tens of millions of dollars to implement AI-based planning and operation of RES. It marks 2025 as a crucial year in the evolution of AI for energy [23]. Globally, the broad adoption of AI in the industry highlights the significant potential of AI to improve energy production, distribution, and utilization further.

Hence, the comparison of AI's prospects and growing potential is discussed in Table 9.1, and the critical domains that require a focus, including quantum

TABLE 9.1

Future Prospective and Growing Domains

Domains	Current Stage	Future Goals	Importance
Quantum Computing	Exploration in the grid-based optimization	Algorithms for Ultra-efficient Optimization	Managing complex decision-making in real-time
Blockchain	Ongoing trial for Peer-to-Peer trading	Decentralized energy market	Secure and transparent transactions for DERs
Autonomous Communities	Decentralized energy sharing	Autonomous AI-based microgrid	Energy independence, community-level support, and decentralization
Edge AI	Initial stages for edge-based real-time management	Broad adoption of Edge AI-based decentralized	Critical for the low-latency decision and operational decisions
Explainable AI	Establishing guidelines for transparency	Integrate the ethical model and guidelines for the energy system	Essential for user confidence, adherence to regulations, and safe utilization

computing and blockchain for grid optimization and energy decentralization, along with the autonomous communities to attain the goal of long-term viability. In addition, edge AI is essential for fast decision support, but explainable AI offers a fair and accountable implementation. Integration of these technologies provides strong support for future development based on an AI-driven EMS system for fair and transparent policy-making in the energy sector.

9.7 Conclusion

Fast adoption of AI with RES has significant possibilities, enhancing operational adaptability and real-time optimized performance. The achievement of the worldwide long-term viability and carbon-zero emission objective depends on the vast adaptability of RES and its accurate forecasting. The significant technical, social, and economic challenges include data confidentiality, benchmarking drawbacks, and the lack of a qualified workforce. Using cutting-edge technologies such as Edge AI, quantum computing, digital twins, and blockchain calls for multidisciplinary collaboration, well-defined legal legislation, and good governance. Dealing with these challenges will ensure AI's vital contribution to developing a strong, viable, entirely self-sufficient future energy system.

References

1. J. Chatterjee, and N. Dethlefs, "Facilitating a smoother transition to renewable energy with AI," *Patterns*, vol. 3, no. 6, p. 100528, 2022, doi: 10.1016/J.PATTER.2022.100528.
2. H. Liu, and R. Yang, "Artificial intelligence for wind speed forecasting: A review on multi-scale decomposition and intelligent fusion strategies," *Advances in Wind Engineering*, vol. 2, no. 2, p. 100055, 2025, doi: 10.1016/J.AWE.2025.100055.
3. "Google harnesses the power of AI to cut energy use | World Economic Forum." https://www.weforum.org/stories/2016/07/google-harnesses-the-power-of-ai-to-cut-energy-use/ (accessed Mar. 19, 2025).
4. "AI adoption in energy should focus on agility | BCG." https://www.bcg.com/publications/2024/ai-adoption-in-energy (accessed Mar. 19, 2025).
5. "New IBM study data reveals 74% of energy & utility companies surveyed embracing AI." https://newsroom.ibm.com/2024-02-26-New-IBM-Study-Data-Reveals-74-of-Energy-Utility-Companies-Surveyed-Embracing-AI (accessed Mar. 19, 2025).
6. I. International Energy Agency, "World Energy Outlook 2023 executive summary," 2023. [Online]. Available: www.iea.org/terms
7. I. Renewable Energy Agency, Renewable capacity statistics 2023 Statistiques De Capacité Renouvelable 2023 Estadísticas De Capacidad Renovable 2023 about Irena. 2023. [Online]. Available: www.irena.org
8. Y. Zhang, T. Huang, and E. F. Bompard, "Big data analytics in smart grids: A review," *Energy Informatics*, vol. 1, no. 1, pp. 1–24, 2018, doi: 10.1186/S42162-018-0007-5/FIGURES/4.
9. M. A. Amin, A. La Fata, R. Procopio, M. Invernizzi, M. Petronijevic, and I. R. Mitic, "Photovoltaic power nowcasting using decision-trees based algorithms and neural networks," pp. 1–6, Sep. 2024, doi: 10.1109/ICETRAN62308.2024.10645147.
10. M. A. Amin, et al., "Renewable energy maximization for pelagic islands network of microgrids through battery swapping using deep reinforcement learning," *IEEE Access*, vol. 11, pp. 86196–86213, 2023, doi: 10.1109/ACCESS.2023.3302895.
11. K. Ukoba, K. O. Olatunji, E. Adeoye, T. C. Jen, and D. M. Madyira, "Optimizing renewable energy systems through artificial intelligence: Review and future prospects," *Energy and Environment*, 2024, doi: 10.1177/0958305X241256293/ASSET/8D92F57B-7887-4AB6-9318-235015A49E74/ASSETS/IMAGES/LARGE/10.1177_0958305X241256293-FIG7.JPG.
12. R. Kabir, D. Halder, and S. Ray, "Digital twins for IoT-driven energy systems: A survey", doi: 10.1109/ACCESS.2024.3506660.
13. D. Lewis, and J. Oxby, "Energy security and AI," Dec. 2024, doi: 10.58248/PN735.
14. "AI for energy opportunities for a modern grid and clean energy economy," 2024.
15. D. Chen, X. Lin, and Y. Qiao, "Perspectives for artificial intelligence in sustainable energy systems," *Energy*, vol. 318, 2025, doi: 10.1016/j.energy.2025.134711.
16. M. S. S. Danish, "AI in energy: Overcoming unforeseen obstacles," *AI*, vol. 4, pp. 406–425, 2023, doi: 10.3390/ai4020022.
17. "AI's energy dilemma: Challenges, opportunities, and a path forward | World Economic Forum." https://www.weforum.org/stories/2025/01/ai-energy-dilemma-challenges-opportunities-and-path-forward/ (accessed Apr. 04, 2025).

18. O. A. Omitaomu, and M. Sayed Shah Danish, "AI in energy: Overcoming unforeseen obstacles," *AI 2023*, vol. 4, no. 2, pp. 406–425, 2023, doi: 10.3390/AI4020022.
19. M. N. Fekri, K. Grolinger, and S. Mir, "Asynchronous adaptive federated learning for distributed load forecasting with smart meter data," *International Journal of Electrical Power and Energy Systems*, vol. 153, p. 109285, 2023, doi: 10.1016/J.IJEPES.2023.109285.
20. S. L. Choi, et al., "Generative AI for power grid operations," 2024, Accessed: Apr. 02, 2025. [Online]. Available: https://www.nrel.gov/docs/fy25osti/91176.pdf.
21. "IEEE SA – Four Energy Trends to Watch for in 2025." https://standards.ieee.org/beyond-standards/2025-energy-trends/ (accessed Apr. 02, 2025).
22. "Artificial intelligence act and regulatory sandboxes".
23. "DOE offers funding to accelerate interconnection process through utilization of artificial intelligence | American Public Power Association." https://www.publicpower.org/periodical/article/doe-offers-funding-accelerate-interconnection-process-through-utilization-artificial-intelligence (accessed Apr. 02, 2025).

Index